WILD ANIMAL WAYS

Free at last

WILD ANIMAL WAYS

By
ERNEST THOMPSON SETON

Author of
"Wild Animals at Home," "Wild Animals I Have Known,"
"Two Little Savages," "Biography of a Grizzly,"
"Life Histories of Northern Animals,"
"Rolf in the Woods," "The Book of Woodcraft,"
Chief of the Woodcraft League of America

With 200 Drawings by the Author

Garden City   New York
Doubleday, Page & Company
1925

Copyright, 1916, by
ERNEST THOMPSON SETON

ALL RIGHTS RESERVED

PRINTED IN THE UNITED STATES
AT
THE COUNTRY LIFE PRESS, GARDEN CITY, N. Y.

Preface

When I look at the names of the animals whose stories are given here, I feel much as an artist might in looking at sundry portraits of his friends and ideals painted by himself.

Some of these I personally knew. Some are composites, and are merely natural history in story form. Way-atcha, Atalapha, and Foam are of the latter kind.

Foam is an effort to show how the wild things instinctively treat themselves in sickness. They have their herbs, their purges, their sudorifics, their hot and cold baths, their mud baths, their fastings, their water sluicings, their massage, their rest cure, and their sun treatment.

The final scene when the Razor-back utterly defeated the Bear was witnessed and related to me long ago by a Michigan lumberman, whose name I cannot recall. The minor incidents are largely from personal observation of wild hogs in various parts of America. I am in hopes that some will

Preface

see the despised Razor-back in a more friendly light when they realize the strong and wise little soul that lurks behind those blinking eyes.

The Wild Geese is a simple narrative of well-known facts, facts that I observed among the Honkers in my own home park.

Jinney, the bad monkey, I never saw, but I have told her story as it was given to me by my old friend Louis Ohnimus, at one time Director of the Woodward Zoölogical Gardens in San Francisco, California.

Billy and Coaly-Bay are in the main true, and a recent letter from the West gives me new light on the history of the wild horse. The story had just appeared in *Collier's Magazine*, where the writer saw it.

The letter runs as follows:

"January 26, 1916. I, too, knew Coaly-Bay, the glorious creature. He began his struggles in the Bitterroot Mountains of Idaho, left through the Salmon River country straggling tales of his fierce resentment under the yoke, and escaped triumphantly at last to the plains in the south.

"I was sixteen then and it is six years ago.

"Something, however, you failed to record. It is this: that before he escaped from the world of

Preface

spur and lash, the world of compulsion, the world that denies to a horse an end in himself, he came to love one person—me, the woman who petted instead of saddled him, who gave him sugar instead of spurring him, who gloried in him because he dared assert that he belonged to himself. For I, too, was an outlaw.

"When I wandered joyfully through the evergreen labyrinths of the Florence Basin, sniffing like a hare or fox the damp spring smell of the earth, going far down the narrow, rock-walled canyons for the first wild orchids, Coaly-Bay came, too. I did not ride or drive him. He trotted beside me as might a dog. We were pals, equals, fellow rebels. I went with him where he could find the first young meadow grass, and he went with me where grew the first wild strawberries. As together we glimpsed, far below, the green ribbon that was the Salmon River, or saw, far off, the snow attempting to cover the sinister blackness of the Buffalo Hump, we laughed at the stupidity of the world of man, who sought to drive things, to compel things, to master things, breeding hate and viciousness thereby; the stupidity of the world of men who never dreamed of the marvellous power of love!

"But they came between us, these men; and

Preface

when Coaly-Bay broke the leg of one of them, I laughed. That day when they were going to crush his spirit with a bullet, I *hated* them! And when he escaped down those endless labyrinths, which we had threaded together so often, how I gloated! But later I wept, for he had left me to be an outlaw alone.

"Yes, always I shall love the memory of Coaly-Bay. He was a symbol of the eternal spirit of Revolt against the Spur of Oppression. My desire is to be as true to that spirit as he was, to fight the lash and spur, to bleed or starve rather than submit."

I gladly quote this letter because it interprets some others of my friends as well as Coaly-Bay.
New York,
February 27, 1916.

Contents

                                                          PAGE

I. Coaly-Bay, The Outlaw Horse

    The Wilful Beauty 3
    The Bear Bait 9
    His Destined End 12

II. Foam, or The Life and Adventures of a Razor-Backed Hog

    The Mother 19
    Lizette and the Bear 23
    The Foundling 26
    Pig, Duck, and Lamb 27
    Foam as Defender 31
    A Bad Old Bear 35
    The Swamp 38
    Smell-power 39
    The Rattlesnake 41
    Wildwood Medicine 46
    Springtime 50
    Grizel Seeks Her Fortune 52
    The Scratching Post 54
    The Lovers 56

Contents

	PAGE
The Wildcat	57
The Pork-eating Bear	63
Hill Billy Bogue	67
The Hog Warrior and the Hounds	70
Lizette and an Old Friend	72
The Bear Claims Another Victim	75
The Defeat of Hill Billy	76
The Day of Judgment	78

III. Way-Atcha, The Coon-Raccoon of Kilder Creek

The Home-seekers	90
The Home	92
Schooling the Children	94
The Mysterious Warning	98
The Hunters	101
The Wayward Child	104
A Merry Life on the Farm	107
The Ancient Foe	111
The Blessed Hollow Tree	116

IV. Billy, the Dog That Made Good

Silly Billy	123
The Professional Rough	127
The Fiery Furnace and the Gold	134

V. Atalapha, a Winged Brownie

The Twins	143
The Schooling of a Brownie	147

Contents

	PAGE
The Undoing of Little Brother	153
Atalapha's Toilet	156
The Coming of the Bridegrooms	158
The Great Southern Trek	161
Northward, Home Again	165
Wings and Friendships	168
The Winged Tiger and the Unknown Death	172
Atalapha Wounded and Captive	178
The Wings That See	181
Atalapha Meets with Silver-brown	187
The Love Fire	189
The Race With the Swallows	195
Lost on the Water	197
The Remorseless Sea	201
The Brownies of the Blood Royal	204

VI. The Wild Geese of Wyndygoul

The Bugling on the Lake	211
The Fifth Commandment	214
Father or Mother	218

VII. Jinny. The Taming of a Bad Monkey

A Dangerous Brute	225
Jinny Finds a New Life	230
The Soul of a Monkey	235

LIST OF FULL PAGE ILLUSTRATIONS

Free at Last	*Frontispiece*
	FACING PAGE
Foam's early days with his Mother . . .	22
The fight with the Kogar's Bear . . .	80
Way-atcha with his Mother and Brothers hunting in the moonlight	101
Hanging to the Big Bear's face, flapping like a rag, was Silly Billy	136
The portrait of a Brownie	144
Chasing the booming June-bug . . .	150
The flittering Brownie host in the moonlight	160

I
Coaly-Bay, the Outlaw Horse

I
Coaly-Bay, the Outlaw Horse
THE WILFUL BEAUTY

FIVE years ago in the Bitterroot mountains of Idaho there was a beautiful little foal. His coat was bright bay; his legs, mane, and tail were glossy black—coal black and bright bay—so they named him Coaly-bay.

"Coaly-bay" sounds like "Koli-bey," which is an Arab title of nobility, and those who saw the handsome colt, and did not know how he came by the name, thought he must be of Arab blood. No doubt he was, in a faraway sense; just as all our best horses have Arab blood, and once in a while it seems to come out strong and show in every part of the creature, in his frame, his power, and his wild, free roving spirit.

Coaly-Bay, the Outlaw Horse

Coaly-bay loved to race like the wind, he gloried in his speed, his tireless legs, and when careering with the herd of colts they met a fence or ditch, it was as natural to Coaly-bay to overleap it, as it was for the others to sheer off.

So he grew up strong of limb, restless of spirit, and rebellious at any thought of restraint. Even the kindly curb of the hay-yard or the stable was unwelcome, and he soon showed that he would rather stand out all night in a driving storm than be locked in a comfortable stall where he had no vestige of the liberty he loved so well.

He became very clever at dodging the horse wrangler whose job it was to bring the horseherd to the corral. The very sight of that man set Coaly-bay agoing. He became what is known as a "Quit-the-bunch"—that is a horse of such independent mind that he will go his own way the moment he does not like the way of the herd.

So each month the colt became more set on living free, and more cunning in the means he took to win his way. Far down in his soul, too, there must have been a streak of cruelty, for he stuck at nothing and spared no one that seemed to stand between him and his one desire.

When he was three years of age, just in the perfection of his young strength and beauty, his real

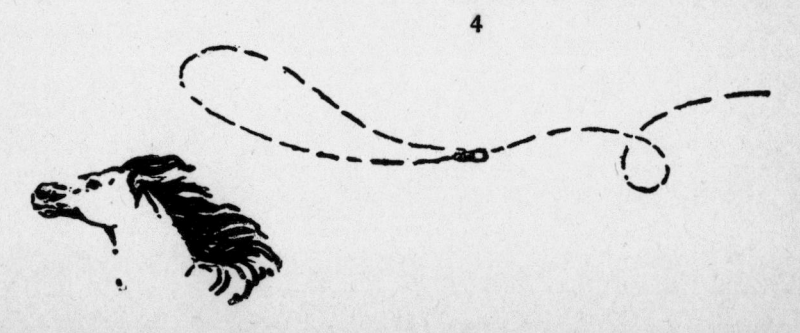

Coaly-Bay, the Outlaw Horse

troubles began, for now his owner undertook to break him to ride. He was as tricky and vicious as he was handsome, and the first day's experience was a terrible battle between the horse-trainer and the beautiful colt.

But the man was skilful. He knew how to apply his power, and all the wild plunging, bucking, rearing, and rolling of the wild one had no desirable result. With all his strength the horse was hopelessly helpless in the hands of the skilful horseman, and Coaly-bay was so far mastered at length that a good rider could use him. But each time the saddle went on, he made a new fight. After a few months of this the colt seemed to realize that it was useless to resist, it simply won for him lashings and spurrings, so he pretended to reform. For a week he was ridden each day and not once did he buck, but on the last day he came home lame.

His owner turned him out to pasture. Three days later he seemed all right; he was caught and saddled. He did not buck, but within five minutes he went lame as before. Again he was turned out to pasture, and after a week, saddled, only to go lame again.

His owner did not know what to think, whether the horse really had a lame leg or was only shamming, but he took the first chance to get rid of him, and though Coaly-bay was easily worth fifty dollars, he sold him for twenty-five. The new owner felt he had a bargain, but after being ridden half a mile Coaly-bay went lame. The rider got off to examine the foot, whereupon Coaly-bay broke away and galloped back to his old pasture. Here he was caught, and the new owner, being neither gentle nor sweet, applied spur without mercy, so that the next twenty miles was covered in less than two hours and no sign of lameness appeared.

Now they were at the ranch of this new owner. Coaly-bay was led from the door of the house to the pasture, limping all the way, and then turned out. He limped over to the other horses. On one side of the pasture was the garden of a neighbor. This man was very proud of his fine vegetables and had put a six-foot fence around the place. Yet the very night after Coaly-bay arrived, certain of the horses got into the garden somehow and did a great deal of damage. But they leaped out before daylight and no one saw them.

The gardener was furious, but the ranchman stoutly maintained that it must have been some other horses, since his were behind a six-foot fence.

Next night it happened again. The ranchman went out very early and saw all his horses in the pasture, with Coaly-bay behind them. His lame-

Coaly-Bay, the Outlaw Horse

ness seemed worse now instead of better. In a few days, however, the horse was seen walking all right, so the ranchman's son caught him and tried to ride him. But this seemed too good a chance to lose; all his old wickedness returned to the horse; the boy was bucked off at once and hurt. The ranchman himself now leaped into the saddle; Coaly-bay bucked for ten minutes, but finding he could not throw the man, he tried to crush his leg against a post, but the rider guarded himself well. Coaly-bay reared and threw himself backward; the rider slipped off, the horse fell, jarring heavily, and before he could rise the man was in the saddle again. The horse now ran away, plunging and bucking; he stopped short, but the rider did not go over his head, so Coaly-bay turned, seized the man's foot in his teeth, and but for heavy blows on the nose would have torn him dreadfully. It was quite clear now that Coaly-bay was an "outlaw"—that is an incurably vicious horse.

The saddle was jerked off, and he was driven, limping, into the pasture.

The raids on the garden continued, and the two men began to quarrel over it. But to prove that his horses were not guilty the ranchman asked the gardener to sit up with him and watch. That night as the moon was brightly shining they saw, not all the horses, but Coaly-bay, walk straight up to the garden fence—no sign of a limp now—easily leap over it, and proceed to gobble the finest things he could find. After they had made sure of his identity, the men ran forward. Coaly-bay cleared the fence like a Deer, lightly raced over the pasture to mix with the horseherd, and when the men came near him he had—oh, such an awful limp.

"That settles it," said the rancher. "He's a fraud, but he's a beauty, and good stuff, too."

"Yes, but it settles who took my garden truck," said the other.

"Wall, I suppose so," was the answer; "but luk a here, neighbor, you ain't lost more'n ten dollars in truck. That horse is easily worth—a hundred. Give me twenty-five dollars, take the horse, an' call it square."

"Not much I will," said the gardener. "I'm out twenty-five dollars' worth of truck; the horse ain't worth a cent more. I take him and call it even."

And so the thing was settled. The ranchman said nothing about Coaly-bay being vicious as well as cunning, but the gardener found out, the very first time he tried to ride him, that the horse was as bad as he was beautiful.

Coaly-Bay, the Outlaw Horse

Next day a sign appeared on the gardener's gate:

> FOR SALE
> First-class horse, sound and gentle. $10.00

THE BEAR BAIT

Now at this time a band of hunters came riding by. There were three mountaineers, two men from the city, and the writer of this story. The city men were going to hunt Bear. They had guns and everything needed for Bear-hunting, except bait. It is usual to buy some worthless horse or cow, drive it into the mountains where the Bears are, and kill it there. So seeing the sign up, the hunters called to the gardener: "Haven't you got a cheaper horse?"

The gardener replied: "Look at him there, ain't he a beauty? You won't find a cheaper horse if you travel a thousand miles."

"We are looking for an old Bear-bait, and five dollars is our limit," replied the hunter.

Horses were cheap and plentiful in that country; buyers were scarce. The gardener feared that

Coaly-Bay, the Outlaw Horse

Coaly-bay would escape. "Wall, if that's the best you can do, he's yourn."

The hunter handed him five dollars, then said: "Now, stranger, bargain's settled. Will you tell me why you sell this fine horse for five dollars?"

"Mighty simple. He can't be rode. He's dead lame when he's going your way and sound as a dollar going his own; no fence in the country can hold him; he's a dangerous outlaw. He's wickeder nor old Nick."

"Well, he's an almighty handsome Bear-bait," and the hunters rode on.

Coaly-bay was driven with the packhorses, and limped dreadfully on the trail. Once or twice he tried to go back, but he was easily turned by the men behind him. His limp grew worse, and toward night it was painful to see him.

The leading guide remarked: "That thar limp ain't no fake. He's got some deep-seated trouble."

Day after day the hunters rode farther into the mountains, driving the horses along and hobbling them at night. Coaly-bay went with the rest, limping along, tossing his head and his long splendid mane at every step. One of the hunters tried to ride him and nearly lost his life, for the horse seemed possessed of a demon as soon as the man was on his back.

Coaly-Bay, the Outlaw Horse

The road grew harder as it rose. A very bad bog had to be crossed one day. Several horses were mired in it, and as the men rushed to the rescue, Coaly-bay saw his chance of escape. He wheeled in a moment and turned himself from a limping, low-headed, sorry, bad-eyed creature into a high-spirited horse. Head and tail aloft now, shaking their black streamers in the wind, he gave a joyous neigh, and, without a trace of lameness, dashed for his home one hundred miles away, threading each narrow trail with perfect certainty, though he had seen them but once before, and in a few minutes he had steamed away from their sight.

The men were furious, but one of them, saying not a word, leaped on his horse—to do what? Follow that free ranging racer? Sheer folly. Oh, no!—he knew a better plan. He knew the country. Two miles around by the trail, half a mile by the rough cut-off that he took, was Panther Gap. The runaway must pass through that, and Coaly-bay raced down the trail to find the guide below awaiting him. Tossing his head with anger, he wheeled on up the trail again, and within a few yards recovered his monotonous limp and his evil expression. He was driven into camp, and there he vented his rage by kicking in the ribs of a harmless little packhorse.

Coaly-Bay, the Outlaw Horse

HIS DESTINED END

This was Bear country, and the hunters resolved to end his dangerous pranks and make him useful for once. They dared not catch him, it was not really safe to go near him, but two of the guides drove him to a distant glade where Bears abounded. A thrill of pity came over me as I saw that beautiful untamable creature going away with his imitation limp.

"Ain't you coming along?" called the guide.

"No, I don't want to see him die," was the answer. Then as the tossing head was disappearing I called: "Say, fellows, I wish you would bring me that mane and tail when you come back!"

Fifteen minutes later a distant rifle crack was heard, and in my mind's eye I saw that proud head and those superb limbs, robbed of their sustaining indomitable spirit, falling flat and limp—to suffer the unsightly end of fleshly things. Poor Coaly-bay; he would not bear the yoke. Rebellious to the end, he had fought against the fate of all his kind. It seemed to me the spirit of an Eagle or a Wolf it was that dwelt behind those full bright eyes —that ordered all his wayward life.

I tried to put the tragic finish out of mind, and

Coaly-Bay, the Outlaw Horse

had not long to battle with the thought; not even one short hour, for the men came back.

Down the long trail to the west they had driven him; there was no chance for him to turn aside. He must go on, and the men behind felt safe in that.

Farther away from his old home on the Bitterroot River he had gone each time he journeyed. And now he had passed the high divide and was keeping the narrow trail that leads to the valley of Bears and on to Salmon River, and still away to the open wild Columbian Plains, limping sadly as though he knew. His glossy hide flashed back the golden sunlight, still richer than it fell, and the men behind followed like hangmen in the death train of a nobleman condemned—down the narrow trail till it opened into a little beaver meadow, with rank rich grass, a lovely mountain stream and winding Bear paths up and down the waterside.

"Guess this'll do," said the older man. "Well, here goes for a sure death or a clean miss," said the other confidently, and, waiting till the limper was out in the middle of the meadow, he gave a short, sharp whistle. Instantly Coaly-bay was alert. He swung and faced his tormentors, his noble head erect, his nostrils flaring; a picture of horse beauty —yes, of horse perfection.

Coaly-Bay, the Outlaw Horse

The rifle was levelled, the very brain its mark, just on the cross line of the eyes and ears, that meant sure—sudden, painless death.

The rifle cracked. The great horse wheeled and dashed away. It was sudden death or miss—and the marksman *missed*.

Away went the wild horse at his famous best, not for his eastern home, but down the unknown western trail, away and away; the pine woods hid him from the view, and left behind was the rifleman vainly trying to force the empty cartridge from his gun.

Down that trail with an inborn certainty he went, and on through the pines, then leaped a great bog, and splashed an hour later through the limpid Clearwater and on, responsive to some unknown guide that subtly called him from the farther west. And so he went till the dwindling pines gave place to scrubby cedars and these in turn were mixed with sage, and onward still, till the faraway flat plains of Salmon River were about him, and ever on, tireless as it seemed, he went, and crossed the canyon of the mighty Snake, and up again to the high wild plains where the wire fence still is not, and on, beyond the Buffalo Hump, till moving specks on the far horizon caught his eager eyes, and coming on and near, they moved and rushed

Coaly-Bay, the Outlaw Horse

aside to wheel and face about. He lifted up his voice and called to them, the long shrill neigh of his kindred when they bugled to each other on the far Chaldean plain; and back their answer came. This way and that they wheeled and sped and caracoled, and Coaly-bay drew nearer, called and gave the countersigns his kindred know, till this they were assured—he was their kind, he was of the wild free blood that man had never tamed. And when the night came down on the purpling plain his place was in the herd as one who after many a long hard journey in the dark had found his home.

There you may see him yet, for still his strength endures, and his beauty is not less. The riders tell me they have seen him many times by Cedra. He is swift and strong among the swift ones, but it is that flowing mane and tail that mark him chiefly from afar.

There on the wild free plains of sage he lives: the stormwind smites his glossy coat at night and the winter snows are driven hard on him at times; the Wolves are there to harry all the weak ones of the herd, and in the spring the mighty Grizzly, too, may come to claim his toll. There are no luscious pastures made by man, no grain-foods; nothing but the wild hard hay, the wind and the open plains, but here at last he found the thing he craved—the one worth all the rest. Long may he roam—this is my wish, and this—that I may see him once again in all the glory of his speed with his black mane on the wind, the spur-galls gone from his flanks, and in his eye the blazing light that grew in his far-off forebears' eyes as they spurned Arabian plains to leave behind the racing wild beast and the fleet gazelle—yes, too, the driving sandstorm that overwhelmed the rest, but strove in vain on the dusty wake of the Desert's highest born.

II
Foam, or The Life and Adventures of a Razor-Backed Hog

II
Foam, or The Life and Adventures of a Razor-Backed Hog

THE MOTHER

SHE was just an ordinary Razor-backed Hog in the woods of South Virginia, long-legged and long-snouted, strong in shoulder, hard and tight in the flanks, and equipped with sharp white tusks that, though short, were long enough to inspire terror in any dog that dared to try her mettle. She roamed in the glades by Prunty's during summer, or in winter, when food was scarce, rendered a half-hearted and mercenary allegiance to the Prunty barnyard which furnished a sort of mart, where many different races met to profit by the garnered stores or waste.

The early spring had passed. Bright summer had begun; redbird and robin were stating it in set

Foam—A Razor-Backed Hog

terms, while wind-root and Mayflower were posting the fact on their low banks, and the Razor-back wandered from under the barn, blinking her pale-lashed eyes. Pensively nosing the ground, she passed by untouched some corn that she certainly smelled, and, a day before, would have gobbled. But she was uneasy and nosed about till she reached the "branch" where she drank deeply. Still swinging slowly, she crossed the stream, and wandered into the woods. She listened hard, and looked back once or twice, then changed her course, crossed the brook twice more—yes, that is their way when they shun pursuit—and wandered on till, far in the shades, she reached an upturned tree root. She had been there before, and the layer of grass and leaves showed the beginnings of a bed. After sniffing it over, she set about gathering more grass, stopping like a statue occasionally when some strange sound was wind-borne to her ears. Once or twice she moved away, but each time returned to lie down uneasily in the nest she had prepared.

Oh Mother, All-mother Nature that lays such heavy hand upon maternity in towns, where help is near! How kind thou art to the wildwood beast that all alone must face the ordeal. How doubly blest is she, in strength and soon deliver-

Foam—A Razor-Backed Hog

ance! And when the morning sun arose, it peeped a rosy peep for a moment under the old gnarled roof-root, to see a brood of cowering pink-nosed piglets, with their mother lying as a living barrier against the outside world.

Young life is always beautiful. And those who picture pigs as evil passions, dirt and lust expressed in flesh would have marvelled to see the baby beauty of that brood and the sweet perfection of the mother's love. She had no eyes for the pretty rounded forms or soft clear tints, but she loved them with her full returning force, and when, with their growing strength and need for food, they nosed and nudged and mouthed her body for their natural sustenance, that double row of noselets gave double thrills of mother joy and dear content. During the time when they could not follow, she grudged the moments when she must slip away to find the needful food and drink, nor went beyond the reach of their slightest call.

Her life all winter had centred in the barnyard. But the wish to keep her young ones hidden made her lead them deeper into the woods when they began to run. And the sportive, rollicking crew, boring their little gimlet noses into everything near and soft, soon grew in vigor and acquired a wonderful knowledge of woodland smells. There were hosts of things to eat in the Maytime woods. Every little early flower has a bulbous root that is a store of food. Every berry that follows the flower is food. And when it so falls out that these be poisonous, and such there be, the good All-mother has put in it a nasty little smell, a funny tang, or a prickle that sounds a warning to the wood-wise pig and makes it unpleasant to the ever-moving finger-tipped inquiring noses of the rollicking grunting piggy band. These were the things the mother knew. These were the things the young ones learned by watching and smelling. One of them, a lively youngster in reddish hair, found a new sensation. They were not eating yet, but the mother was rooting and eating all day, and the youngsters rushed to smell each new place that she upheaved. Grubs she welcomed as a superior kind of roots, and the children sniffed approval. Then a queer, broad, yellow-banded, humming, flying thing dropped down on a leaf near Redhead's nose. He poked it with his nose finger-tip. And then it did—it did—something he could not understand, but oh, how it hurt! He gave a little "Wowk" and ran to his mother. His tiny bristles stood up and he chopped his little foxlike jaws till they foamed, and the white froth flecked his cheeks. It was a sun and night before little Foamy Chops had

Foam's early days with his Mother

got over it, but it did him no serious harm, and he remembered.

They had been running a week or more in the woods when something happened to show how the mother's mind was changed by her family. Loud rumbling noises were heard not far ahead, and now they were coming near. Mother understood them quite well—the sounds of men approaching. She had long known such sounds in the barnyard days as promise of food, but now she thought of her brood. It might mean danger to them, and she turned about, giving a low "Woof" that somehow struck terror into the hearts of the young ones. They had never heard that before, and when she wheeled and walked quickly away, the brood went scrambling behind her in a long silent troop, with Foamy Chops at his mother's tail.

This was a small incident, but it was a turning point, for thenceforth the mother and her brood had broken with the barnyard and its folk.

LIZETTE AND THE BEAR

Lizette Prunty was a big girl now, she was thirteen and not afraid to go far alone in the hills. June with its sweet alluring strawberries was in the woods, and Lizette went afield. How is it that the berries just ahead are always bigger, riper, and more

Foam—A Razor-Backed Hog

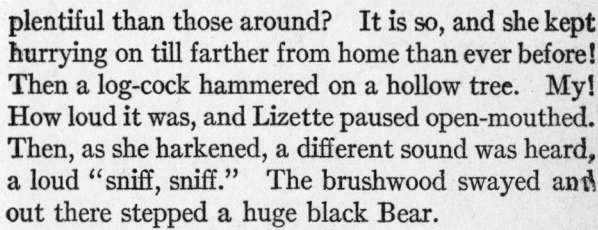

plentiful than those around? It is so, and she kept hurrying on till farther from home than ever before! Then a log-cock hammered on a hollow tree. My! How loud it was, and Lizette paused open-mouthed. Then, as she harkened, a different sound was heard, a loud "sniff, sniff." The brushwood swayed and out there stepped a huge black Bear.

At the little frightened "Oh!" the Bear stopped, reared up to his great height, and stood there gazing and letting off, at each few seconds, a loud, far-reaching "Woof." Poor Lizette was terror stricken. She could neither speak nor run. She simply stood and gazed. So did the Bear.

Then another noise arose, a deep grunt and a lot of little grunties. "A whole pack of Bears," thought poor Lizette, but she could not move. She merely gazed toward the new sounds. So did the Bear.

This time when the tall grass parted it was to show, not a lot of Bears, but the old Razor-back long missing from the barnyard, and her lively grunting brood.

Very rarely does a Bear molest a child, very rarely does he miss a chance for pork. The black monster dropped on all fours and charged at the mother and her brood.

The fierce defiant war-grunts of the mother might

Foam—A Razor-Backed Hog

have struck terror into any but a big black Bear, for the Razor-back had sharp tusks and mighty jaws, and sturdy legs, and flanks all armored well with double hide and bristle thatch, and—the heart of a devoted mother.

She stood her ground and faced the foe, while the little ones, uttering cries of fear, crowded against her sides or hid behind her. Only little Foamy stood with his head aloft to watch the awful enemy.

Even a Bear must be impressed when a Razor-back is out in fighting mood to save her young, and he walked around the group while she ever turned to face him. She had backed into a protecting bush that made any but front attack impossible. And the Bear walked this way and that, without seeing any good chance to close, for the mother always fronted him, and those champing armed jaws were not to be lightly faced.

Then the Bear made a short charge and stopped. The mother, ever fronting, saw him pause, and now she charged. She ripped his arm and bit the other paw, but he was on her now, and in a rough and tumble the Bear had every chance. He stunned her with a blow, he raked her sides, he crunched her leg. He gripped her in a fierce embrace that robbed her of all fighting breath, while his hind claws

Foam—A Razor-Backed Hog

ripped her open, and as they struggled in the final throe Lizette recovered use of sense and limb; she turned and fled for home.

THE FOUNDLING

"Oh, father it was awful! Just down by Kogar's Creek. I can take you there in half an hour."

So father came with dog and gun. Lizette was guide, and in a little while they were among the strawberry tracts of Kogar's Creek. Turkey-buzzards were sailing over the place as they drew near. They found the very spot. There lay the mother Razor-back, torn and partly devoured. Under her body and half hidden about were the young, crushed, each of them, by one blow of that cruel mighty paw.

Prunty was uttering mannish grunts and growls at each fresh discovery, Lizette was weeping, when the dog broke into a tirade at something far under the bush; and bravely facing him there showed a little red-headed piglet, chopping with his tiny jaws till the foam flew, and squeaking out defiance to the new terror.

"Hello, there's one escaped!" exclaimed father. "Isn't he sassy?" So while little Foamy was heroically facing the dog, the father reached through the brush from behind, and seizing the

Foam—A Razor-Backed Hog

piggie by the hind leg, he lifted him protesting, squealing, and champing, to drop him into his game bag.

"Poor little chap, see how his nose is skinned! He must be hungry. I'm afraid he's too young to live."

"Oh, do let me have him, Father; I'll feed him," and so Lizette's moral claim to Foamy was legalized on the spot.

Prunty had brought a huge bear trap to the place, and now he set it by the body of the victim. But all it ever caught there was an unlucky turkey-buzzard. The Kogar's Creek Bear was too cunning to be taken by such means: and buzzards, insects, and kindly flowers wiped out all tragic records on that spot.

PIG, DUCK, AND LAMB

Poor little Foamy Chops. He was so hungry, so forlorn, and his nose was so sore where the Bear had scratched him. He did not know that Lizette was his friend, and he champed his little harmless jaws at her in defiance when she put him in the box that was to take the place of all outdoors for him. She washed his wounded nose. She brought him some warm milk in a saucer, but he did not understand it that way. Hours went by and still he

Foam—A Razor-Backed Hog

crouched in dull, motionless despair. Then Lizette's own nurse came with a feeding bottle. Foam kicked, squealed, and champed his jaws, but strong hands wrapped him up in a cloth. The bottle feeder was put to his open mouth. It was warm and sweet. He was oh! so hungry now! He could no more help sucking than any other baby could, and when the bottle was empty, he slept the long sweet sleep he so much needed.

When you help some one it always makes you love that some one very much; so of course Lizette was now devoted to little Foam; but he knew her only as a big dangerous thing, and hated her. Yet not for long. He was an intelligent little Razorback; and before his tail had the beginning of a curl he learned that "Lizette" meant "food," so he rose each time to meet her. Next he found he could bring Lizette—that is, food—if he squealed, and thenceforth his daily practice developed a mighty voice.

In a week his shyness was gone. He was now transferred to a stall in the stable. In a month he was tame as a cat and loved to have his back scratched, and the large wound on his nose was healed, though it left an ugly scar.

Then two companions entered his life, a duck and a lamb, strange creatures that Foam inspected nar-

Foam—A Razor-Backed Hog

rowly out of his white-rimmed eyes, with distrust and a little jealousy. But they proved pleasant persons to sleep with; they kept him so warm. And soon he devised means of enjoying them as playthings; for the lamb's tail was long and pullable, and the duck could be tossed over his back by a well-timed "root!"

The box stall was now too small, but a fenced-in yard gave ample runway. Here in the tall weeds little Foam would root and race, or tease his playmates, or hide from his foster-mother. Yes, many a time when she came and called she had no response; then carefully, anxiously searching about she would come on the little rascal hiding behind some weeds. Knowing now that he was discovered, he would dash forth grunting hilariously at every bound, circling about like a puppy, dodging away when she tried to touch him, but at last when tired of the flirtation he would surrender on the understanding that his back was to be scratched.

Many a circus has shown the wondering world a learned pig, a creature of super-animal intelligence, and yet we say of a dull person, "He is as stupid as a pig," which proves merely that pigs vary vastly. Many are stupid, but there are great possibilities in the race; some may be in the very front rank of animal intelligence. The lowest in the scale of

Foam—A Razor-Backed Hog

pigs is the fat porker of the thoroughbred farm. The highest is the wild Razor-back, who lives by his wits. And soon it was clear that Foam was high in his class. He was a very brainy little pig. But he developed also a sense of humor, and a real affection for Lizette.

At the shrill whistle which her father had taught her to make with her fingers in her teeth, he would come racing across the garden—that is, he would come, unless that happened to be his funny day, when, out of sheer caprice, he would hide and watch the search.

One day Lizette was blacking her shoes with some wonderful French polish that dried quite shiny. It happened to be Foam's day to seek for unusual notice. He tumbled the lamb on top of the duck, ran three times around Lizette, then raised himself on his hind legs and put both front feet on the chair beside Lizette's foot, uttering meanwhile a short whining grunt which was his way of saying, "Please give me some!" Then Lizette responded in an unexpected way: she painted his front feet with the French blacking, which dried in a minute, and Foam's pale pinky hoofs were made a splendid shining black. The operation had been pleasantly ticklesome, and Foam blinked his eyes, but did not move till it was over.

Foam—A Razor-Backed Hog

Then he gravely smelled his right foot, and his left foot, and grunted again. It was all new to him, and he didn't just know what to make of it; but he let it pass. It was not long before the wear and tear of his wearing, tearsome life spoiled all his French polish, and next time Lizette got out her brush and blacking Foam was there to sniff that queer smell and offer his hoofs again for treatment. The sensation must have pleased him, for he gravely stood till the operation was done, and thenceforth every blackening time he came and held his feet for their morning shine.

FOAM AS DEFENDER

Has a pig a conscience? What do you mean by conscience? If it means a realization that one is breaking a law, and that it will bring punishment and that a continuation will surely pile up harder punishment, then animals have consciences in proportion to their brains. And Foam, being born with ample wits, had judge and jury, accuser and witness, in his own heart when he himself was criminal.

He had been forbidden to tease the lamb, who was a harmless woolly fool, and the duck, who was worse. Scolding and switching were things he understood, and because they were finally associated with teasing his companions, he learned that the last delightful pleasures must be classed as crime. More than once when he was riotously chasing Muff or tumbling Fluff into the buttermilk, his mistress, without showing herself or speaking, merely gave a short whistle, the effect of which was to send a guilty-looking little pig to hide in the bushes. Surely he was conscience-stricken.

Now it happened one morning that Lizette looked from her window over the garden and saw Foam standing very still, with his head low and sidewise, his eyes blinking, the very tip of his tail alone twisting—just his attitude when planning some mischief. She was about to use her whistle, but waited a moment to be sure. The lamb was lying under the tiny rainshed in a sort of dull somnolence. Suddenly the duck said "Quack," and ran from the grass to cower beside the lamb. The latter gave a start and blew its nose. Then out of the tall weeds there dashed a lumbering, wolfish puppy dog, breaking into a volley of glorious "yaps" as he charged on the helpless duckling. What fun it was! And the lamb, too, was so frightened that the valiant puppy assailed it without fear.

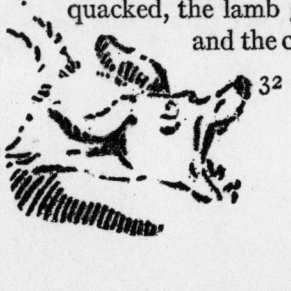

"Yap, yap, yap!" How brave a dog can be when his victim runs or is helpless! The duck quacked, the lamb gave a bleat of terror, and the cur, intoxicated by success and hankering for the highest glories known to his kind, rushed on the duckling, tore off mouthful after mouthful of feathers from his back, and would in a little while have rended him in pieces. But another sound was heard, the short hoarse "Gruff, gruff, gruff" sounds that mean a warpath pig. We call them grunts, because made by a pig, but the very same sounds uttered by a Leopard are called short roars, and these were what came naturally from Foam as he bounded into the scene. Every bristle on his back was erect, his little eyes were twinkling with green light. His jaws, now armed with small but sharp and growing tusks, were chopping the malignant "chop, chop" that flecks the face with foam, proclaims the warlust, and lets the wise ones know that the slumbering wild beast deep inside is roused. Not love of the duck, I fear, but the urge of deep-laid ancient hate of the Wolf, was on him: "a Wolf was raiding his home place." The spirit of a valiant battling race was peeping from those steadfast eyes. Race memories of ancestral fights boiled in his blood. Foam charged the dog.

Was ever bully more surprised? Gleefully the puppy had clutched the duckling's wing to drag him forth, when the little avalanche of red rage pig was on him, and the heave that struck his ribs had

Foam—A Razor-Backed Hog

pins in it; it tumbled him heels over head, scratched and even bleeding. His yaps of glorious victory were changed into howls and yelps of dire defeat. Foam was on him again. The cur sought to escape; limping, howling through a mouthful of plundered feathers, he raced around the shed with Foam behind, then out the door, and through the weeds. A cur with a tail all tin-bedecked went never more loudly or more fast, and where or how he cleared the fence was almost overquick for certain seeing, and whence he came, or whither he went, was far from sure—only this: that his yelping died away in the woods and no more was seen of him.

Lizette and her father both were on hand. Their dumb astonishment at the unexpected quality discovered in the little Razor-back was followed by wild hilarity at the discomfort of the cur, and his ignominious flight before the roused and valorous Foam.

They went into the garden, and the pig came running to them. Lizette was a little in awe of him at first, but he was now no longer a fighting demon, just a funny rollicking little Razor-back, and when she wondered what he would do next, and what she should do, he held up both his feet on a bench that she might give them their morning coat of polish, and stuck his nose so tight between them that she gave that a coat of blacking, too.

Foam—A Razor-Backed Hog

Lizette maintains that Foam ceased teasing the lamb and the duck from that time. He certainly ceased soon after, for the duck was grown up and soon waddled off to join his web-footed kinsmen on the pool, and he and the lamb parted company in an unexpected manner.

A BAD OLD BEAR

Just as there are rogues among Elephants, idlers among Beavers, and mangy man-eaters among Tigers, so there are outlaws among Bears—creatures at war with all the world; perverted brutes that find pleasure chiefly in destruction, making themselves known by their evil deeds, and in the end making enemies strong enough to turn and rend them. The Kogar's Creek Bear was one of these cruel ones. So far as any one knows he never had any family of his own, but roamed into the Kogar's Creek woods probably because his own kind drove him out of their own country in the mountains. So he drifted into Mayo Valley, where Bears were scarce, and wandered about doing all the mischief he could, smashing down fences, little sheds, or field crops that he could not eat, for the pleasure of destroying. Most Bears eat chiefly vegetable food, preferring berries and roots; some Bears eat a little of all kinds, but Kogar's had such a perverted taste that all he sought was flesh. Calf's flesh he loved, but he would not dream of facing a cow, much less a bull. He delighted in robbing birds' nests, because it was so easy: he would work half a day at a hole to get at a family of Flying Squirrels. At first almost any kind of flesh suited him; and he had eaten more than one little baby Bear that chanced to stray from its mother. But his favorite food was pork. He would go a long way for a porker, and when he caught it, he would keep it alive as long as possible for the pleasure of hearing it squeal.

Of course he took only little ones that were unprotected, and it was a great surprise to him that day when Foam's mother made such a fight. He had always thought that pigs of that size were easy game. He took revenge on the little ones, and he growled and limped for many a day after the affair. It kept him away from Razor-backs and he preyed on little Rabbits in their nests, and such things as could not defend themselves. But his wounds healed, he forgot the lesson of that day, and longed for a feast of pork.

A wonderfully keen nose had the Kogar's Bear. The wind was a wireless laden with stories for him, and it needed but a little study to discover some special message, then a following up to reap the benefit.

Foam—A Razor-Backed Hog

He was not far from Prunty's when the soft breeze rippling through the dawn woods brought to him the sweet alluring smell of pig, and he followed it, swinging his black head as he sifted out the invisible trail from others on the wind.

Marvellously silent is a Bear going through the woods, the biggest, bulkiest of them pass like shadows, and Kogar's reached the Prunty homestead swiftly and noiselessly, led at last to the little paddock where Foam, the author of the guiding smell, was sleeping with his head across the woolly back of the lamb.

After a brief survey of the fence the Bear, finding no opening, proceeded to climb over. But it was not meant for such a bulk of flesh; the paling swayed, yielded, and fell, and the Bear was in the paddock.

If Foam had been slower, or the lamb had been quicker, everything would have been different. The Bear rushed forward, Foam darted aside, the lamb sat still, and a heavy blow from the Bear's paw put an end to its chance of ever moving just as Foam disappeared through the hole in the fence and was lost to sight in the thicket.

The Bear's march was soundless indeed, but the crack of the fence, the bleat of the lamb, the rush of that charge, the scared but defiant *snort, snort, snort* of Foam as he rushed away, made noise enough

Foam—A Razor-Backed Hog

to rouse the farmhouse, for it was in truth just on their rousing time, and the farmer peered forth to see a big black Bear scramble over the fence with the lamb in his jaws.

Then was there a great noise, shouting for dogs, holloaing for men, and Prunty, with the ready rifle in hand, dashed into the woods after the Bear.

How slowly a caged Bear seems to hulk around, how little does it let us know the speed of a wild, free Bear on rugged ground. The brambles, rocks, and benches seemed designed to hinder the dogs, but the Bear passed swiftly on. Then the broad expanse of Kogar's Creek was reached, the Bear launched forth to swim across. The strong stream bore him swiftly down. It was pleasant to ride the flood and see the banks go slipping behind him, so lazily he rode, till the hounds' loud baying was faint in the distance, before he paddled out on the other side. And the dogs when they came to the spot were baffled, nor did a search of the other bank shed any light on the mystery.

Far back on the trail they found the body of the lamb.

THE SWAMP

It was sport for the men and fierce joy for the dogs. Lizette alone seemed to suffer all the horror

Foam—A Razor-Backed Hog

and loss. She searched the little paddock in vain, then whistled and whistled.

She followed the trail of the hunters as far as she could, and then at the edge of a thick swamp she stopped. She was all alone. The swamp was open water or mud; it seemed foolish to go on, so she listened a minute, then gave two or three sharp whistled blasts. A soggy noise was heard, a splashing that gave her the creeps, it sounded so Bearlike. Then a grunt, and there appeared a muddy beast of no particular shape, but surely at one end were two small blinking eyes and from somewhere beneath them a friendly sounding grunt. Yes, surely it was, no—yes, now she was sure, for the wanderer had shaken off most of the mud and was upreared, holding his two forefeet on the log to have his hoofs polished; and they needed it as never before, nor was he quite content till Lizette had taken a stick and carried out their ancient understanding by scratching his muddy back.

SMELL-POWER

Only the animal man with a nose can understand the masterfulness of smells, how through the memory they can dominate the brain, and without regard to the smell itself or anything but the memories, be things of joy or pain or fear. Foam had nearly forgotten his early days and his mother's death, but his nose had not, and the smell of Bear had brought it back, and driven him forth in a terror stampede.

That was why he had heard without heeding the old, familiar whistle call.

But the fear was over now; therein lies courage, not to be without fear, but to overcome it. And Foam rioted around, circling full tilt through the bushes around Lizette, stopping short and stock-still in the pathway, head down, eyes twinkling, till Lizette made a pass at him with a stick. Then away he went, careering, pirouetting, and snorting the little joy snorts that in pig talk stand for "Ha! ha! ha!"

Thus they neared the house, when all at once the merry pig was gone. Foam stood like a pointer at a certain spot. His bristles rose, his eyes snapped green, and his jaws, well armed already, were champing till they foamed. Lizette came near to stroke him; he stepped aside, still champing, and now she saw and understood: they were crossing the fresh trail of the Bear; that terrible odour was on it.

But—and this escaped Lizette at the time—the actions of Foam now no longer told of fear; that he had overcome: this pose, his deep-voiced "woof," his menacing tusks, his green-lit eyes, though he

Foam—A Razor-Backed Hog

was but half grown, were the signs of a fighting Boar. She little guessed how much the spirit in him yet might mean to her. Yes, ere two moons had waned her very life indeed was doomed in absence of all human help to rest in keeping of that valiant little beast, protected only by the two small ivory knives he bore, and the heart that never found in fear its guide.

THE RATTLESNAKE

October is summer still in South Virginia; summer with just a small poetic touch of red-leaf time, and Lizette, full of romantic dreams, with little daring hopes of some adventure, too, had gone up the Kogar's Creek to a lonely place to swim in the sluggish bend. She was safe from any intrusion, so did not hesitate to strip and plunge, rejoicing in the cooling water, as only youth in perfect health can do when set in a perfect time. Then she swam to the central sandbar and dug her pink toes into the sand as she courted the searching sunbeams on her back.

Satisfied at length, she plunged to swim across to the low point that was the only landing place, and served as a dressing-room. She was halfway over when she saw a sight that chilled her blood. There coiled on her snowy clothes with head upright, regardant, menacing, was a Banded Rattlesnake, the terror of the mountains, at home in woods or on the water.

It was with sinking heart and trembling limbs that Lizette swam back and landed again on the sandbar.

Now what? A boy would have sought for stones and pelted the reptile away, but there were no stones, and if there had been, Lizette could not throw like a boy.

She did not dare to call for help, she did not know who might come, and she sat in growing misery and fear. An hour dragged slowly by, and the reptile kept its place. She was roasting in the sun, the torment of sunburn was setting in. She must do something. If only father would come! There was just a chance that he might hear her whistle. She put her fingers in her teeth and sent forth the blast that many a Southern woman has had to learn. At first it came out feebly, but again and again, each time louder it sounded, till the distant woods was reached, and she listened in fear and hope. If father heard he would know, and come. She strained her ears to catch some sound responding.

The reptile did not move. Another half-hour passed. The sun was growing fiercer. Again she gave the far-reaching call; and this time, listening, heard sounds of going, of trampling, of coming; then her heart turned sick. Some one was coming. Who? If it were her father he would shout aloud. But this came only with the swish of moving feet. What if it should be one of those half-wild negro tramps! "Oh, father, help!" She tried to hide as the sounds came nearer—hide by burying herself in sand.

The reptile never stirred.

The bushes swayed above the steep bank. Yes, now she saw a dark and moving form. Her first thought was a "Bear." The bushes parted, and forth came little Foam, grown somewhat, but a youngster still. Lizette's heart sank. "Oh, Foam, Foamy, if you only could help me!" and she sent a feeble whistle that was meant for her father, but the Razor-back it was that responded.

Passing quickly along the bank, he came. There was but one way down. It led to the little sandy spit where lay her clothes, and her deadly foe.

Overleaping logs and low brush came the agile Razor-back. He landed on the sand, and suddenly was face to face with the rattling, buzzing banded Death.

Both taken by surprise recoiled, and made ready for attack. Lizette felt a heart clutch, to see her old-time playmate face his fate. The Boar's crest arose, the battle light came in his eyes, the "chop, chop" of his weapons sounded; the age-long, deep instinctive hatred of the reptile came surging up in his little soul, and the battle fire was kindled there, with the courage that never flinches.

Have you heard the short chopping roar that rumbles from the chest of a boar on battle bent—a warcry that well may strike terror into foemen who know the prowess that is there to back its promise? Yes, even when it comes from the half-grown throat of a youngster, with mere thorns for tusks.

In three short raucous coughs that warcry came, and the Boar drew near. His golden mane stood up and gave him double size. His twinkling eyes shone like dull opals as he measured up his foe. He was a little puzzled by the white garments, but edging around for a better footing, he came between the reptile and the stream, and thus, unwittingly, he ended every chance of its escape.

No mother but Mother Nature taught him the moves. Yet she was a perfect teacher. Nothing can elude the Rattler's strike. It baffles the eye; lightning is not swifter. Its poison is death to all small creatures when absorbed, and absorbents there are in every creature, all over its body, except

Foam—A Razor-Backed Hog

on the cheeks and shoulders of a pig. Presenting these then, Foam approached. The Rattler's tail buzzed like a spinner, and his dancing tongue seemed taunting. With a clatter of his ivory knives and a few short, coughlike snorts, the Razor-back replied, and approached guardedly, tempting the snake to strike at its farthest possible range. Both seemed to know the game, although it must have been equally new to both. The snake knew that his life was at stake. His coils grew tighter yet, his baleful eyes were measuring the foe. A feint, and another, and a counter feint, and then—flash, the poison spear was thrown. To be dodged? No, no creature can dodge it. Foam felt it sting his cheek, the dreadful yellow spume was splashed on the wound, but only less quick was his sharp up-jerk. His young tusks caught the reptile's throat and tossed it as he had often tossed the duckling, and ere the poison reptile could recover and recoil, the Razor-back was on him, stamping and snorting. He ripped its belly open, he crushed its head, champing till his face and jaws were frothed, grunting small war-grunts, and rending, nor ceased till all there was left of the death-dealer was evil-smelling rags of scaly flesh ground into the polluted dust.

"Oh, Foam, oh, Foamy, God bless you!" was all

Foam—A Razor-Backed Hog

Lizette could say. She almost fainted for relief. But now the way was clear. A dozen strokes and she was on the point beside the Boar. Una had found her Lion again.

And Foam, she hardly knew what to think of him. He curveted around her on the sand. She almost expected to see him sicken and fall; then joyfully, thankfully she remembered what her father had told her of the terrors of snake-bite, from which the whole hog race was quite immune.

"I wish I knew how to reward you," she said with simple sincerity. Foam knew, and very soon he let her know: all he asked in return was this: "You scratch my back."

WILDWOOD MEDICINE

Are the wild things never ill? Is disease unknown among them? Alas! we know too well that they are tormented pretty much as we are. They have a few remedies that are potent to help the strong, but the weak must quickly die.

And what are the healing things they use? How well they are known to every woodsman! The sunbath, the cold-water bath, the warm-mud bath, the fast, the water cure, the vomit, the purge, the change of diet and place, and the rest cure, with

Foam—A Razor-Backed Hog

tongue massage of the part where there is a bruise or an open wound.

And who is the doctor who prescribes the time and measure? Only this: the craving of the body. Take the thing and so much of it as is agreeable; when it becomes painful or even irksome, that is the body's way of saying "enough."

These are the healing ways of animals, these are the things that every woodsman knows. These are the things that are discovered anew each generation by some prophet of our kind. If he calls them by their simple names he is mocked, but if he gives them Latin names, he is a great scientist and receives world rewards.

Autumn came on Mayo Valley, a thousand little yellow fairy boats were sailing southward on Kogar's Creek, and the "pat, pat, pit" of falling nuts was heard through all the woods. Rich, growing food are nuts, and Foam was busied stuffing himself each day: racing perhaps after butterflies, pretending to root up some big tree, kneeling to swing his head and gash the sod with his growing tusks, springing to his feet to bound a few yards, then halt in a moment, frozen to a statue. Rejoicing in his strength, he grew more strong, and the skating of the final leaves that left the trees found him grown in shank and jaw, lank and light as yet, but framing

Foam—A Razor-Backed Hog

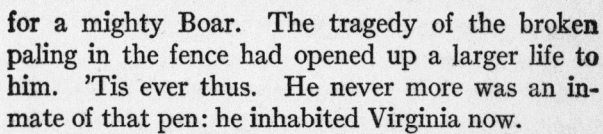

for a mighty Boar. The tragedy of the broken paling in the fence had opened up a larger life to him. 'Tis ever thus. He never more was an inmate of that pen: he inhabited Virginia now.

Down in the black muck swamp he had discovered the trailing ground-nut vines, and when he rooted them out, his nose said, "These are good." Yes, he remembered dimly that his mother used to eat that smell. They furnished a pleasant change from the tree nuts, and he feasted and grew fat. Then he rooted out another old-time root, with a fierce and burning tang, he knew that without munching it, and he tossed the root aside with others of its kind; big, fat, and tempting to the eye they were, but Foam had a safer guide.

Then gorged, he wandered to a sunny slope and, grunting comfortably, dropped flat side flop upon the leaves in lazy, swinish ease.

A bluejay flew just above and shrieked, "You rooter, you rooter!" A wood-pewee snapped flies above his ear, a bog-mouse scrambled over his half-buried leg, yet Foam dozed calmly on.

Then afar a strange sound stirred the silence, a deep-voiced, wailing, whining "Wah-wah-wah, wow-w-w!" then almost screaming, then broken by sobs and snorts, and sometimes falling and muffled, then clear and near—the strangest, mad-

FOAM RAZORBACK Esq.
VIRGINIA
U.S.A
"The Woods"

Foam—A Razor-Backed Hog

dest medley, and so strong it must be the voice of some great forest creature.

Foam was on his feet in a heart-beat, and stock-still there for ten. Now nosing like a pointer with ears acock, with every sense at strain, he crept forward like one spell-drawn.

Slowly back to the rich bottomland the weird sounds led, and then peering through the wire grass he saw his ancient foe, rooting up, crunching, swallowing one after another those terrible burning roots, the white round roots that sting, that tear your very throat, that gripe your bowels, that wring the cheeks with torture like the brands that men leave in the smoking summer land.

Yet on he kept digging, munching, weeping, wailing—digging another, munching it as the tears rolled from his eyes, and the burning pain scorched his slobbering jaws. And still another did that great black monster dig and mouth, and wept and wailed as he did so, and another and another was crowded down his sobbing throat.

Was he insane? Far from it. Was he starving? Not so; the ground was thick with nuts. Then why this dreadful, self-inflicted pain? Who was his master that could order it? Foam had no thoughts about it. The Bear himself could have told you nothing. And yet he was yielding to an

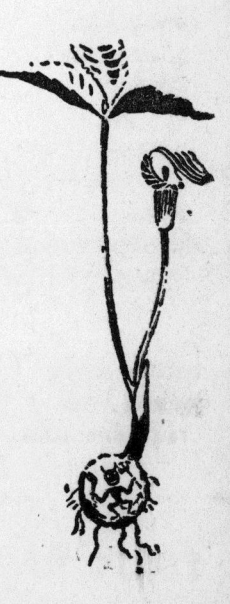

Foam—A Razor-Backed Hog

overmastering inner guide. And these are things we think, but do not surely know: the Bear that seeks only meat for food invites a dire disease that chiefly hurts the skin, and doubly those who make that diet flesh of swine.

It is an ailment of burning skin; the body seems in torment of a myriad tiny fires. And this we think we know: the fiery root affords relief—a slow but sure relief.

And Foam, a youngster yet, afraid, but less afraid, backed slowly from the field a little puzzled, wholly uncomprehending anything but this: his enemy was eating roots and bawling as he ate, and still was bawling out aloud when Foam was far away.

SPRINGTIME

It was a bountiful harvest in the woods that year, and when the branches were bare, the chicaree had seven hollow trees crammed with nuts and acorns, and a well-lined nest near each.

The Muskrat had made huge haycocks in the marsh, the Woodchucks were amazing fat, and every Tree-mouse laid up food as for a three years' famine. The warning of the signs so clear came true: the winter was hard and white.

The woods had been mightily pleasing to young Foam, but now were dull and dreary. His bristly hair grew long and thick as the weather cooled, but not enough; a colder storm set in and Foam at last was forced to seek the shelter of the barn. There were other pigs about, most of them vulgar porkers of the fat and simple table sort, but there were also one or two aristocrats of the real Razor-back strain. At first they were somewhat offish, inclined to thrust him aside like a mere pedigreed pig, but his legs were stout and his tusks were sharp, and he stood quite ready to make good. So by steps he joined himself to the group that snuggled under the barn by night and took its daily comfort at a trough—kinsmen mildly tolerant of each other.

The winter passed and sweet Mistress April of the little leaves was nigh. The influence of the time was on the hills and in the woods; it even reached under the barn among the pigs and stirred them up to life, each in his sort. The fat porkers came slowly forth to the sun, placidly grunting and showing a mild concern in such things of interest as came in range of their low-level vision.

Foam trotted forth like a young colt. How long his legs had grown! How big he was! What shoulders and what a neck of brawn! He was taller than any other in the yard, his gold-red hair was rank, and on his neck and back it made a great hyena mane. When he walked there was spring in his feet, alertness in his poise, and the logy porkers seemed downladen with themselves as they slowly heaved aside to let him pass. The joy of life was on him, and he tossed a heavy trough up in the air, and curveted like a stallion. Then a distant sound made him whirl and run like a mustang. It was Lizette's whistle. They had come very close together that winter, and clearing the low wall like a Deer, Foam reached the door to get a special dish of things he loved, to have his back scratched, and, last, to hold up his forefeet for a rubbing, if not indeed each time for a coat of polish.

"That Foam, as ye call him, Lizette, is more dawg than hawg," Farmer Prunty used to say as he watched the growing Razor-back following the child or playing round her like a puppy—a puppy that weighed 150 pounds, this second springtime of his life. But Foam was merely reviving the ways of his ancestors, long lost in sodden prison pens.

GRIZEL SEEKS HER FORTUNE

It's a long dusty road from Dan River Bridge to Mayo, yet down its whole length there trotted a sleek young Razor-back. She was barely full

Foam—A Razor-Backed Hog

grown, shaped in body and limbs like a Deer, and clad in a close coat of glistening grizzly hair that flashed in the sun when the weather was right, but now was thickly sprinkled with the reddish dust of the old Virginia highway.

Down the long pike she trotted, swinging her sensitive nose, cocking her ears to this or that sound, running some trace a while, like an eager Fox, or making a careful smell study of posts that edged her trail, or marked the trails of offshoot.

An hour, and another hour, she journeyed on, with the steady tireless trot of a searching Razor-back, alert to every promise offered by her senses.

The miles reeled by, she was now in Mayo Valley, but still kept on. Now she found a good rubbing post. It seemed somewhat pleasing to her, she used it well, but soon went on.

What was she doing?

How often we can explain some animal act by looking into ourselves. There comes a time in the life of every man and woman when they are filled with a yearning to go forth into the world and seek their fortune. And the wise say, "Let them go!" This same impulse comes on wild things, and the wise ones go. This, then, is what Grizel was doing. She was seeking her fortune.

Foam—A Razor-Backed Hog

She stopped at many a crossroad and she studied many a faint suggestion on the breeze, but she still kept trotting on, till evening saw her in the woods that lies beyond the lower bridge of Kogar's Creek.

THE SCRATCHING POST

Of all the scratching posts on Prunty's farm quite the best was the rough old cedar corner that marks the farthest point of pasture down the swale. A rough trunk for a rough corner, so it still bore in its imperishable substance the many short knots of its living days. They made a veritable comb at just the fittest height. Every pig in the pasture knew it well. None passed it without a halt to claim its benevolence.

The Prunty swine were loitering near; the huge old grandam shouldered another back so she might rub. Then Foam came striding by. His strength and tusks had weeks past given him right of way. He neared the post. Then, shall I tell it, the post sang out aloud, yes, sang aloud, in a tongue that you or I could never have understood. Even could our duller senses have heard it, what message could we get from:

"Klak-karra, klak-karra
Gorka-li-gorra-wauk?"

Foam—A Razor-Backed Hog

But Foam, whose eyes here helped him not, was all ablaze. Not waiting for the huge old hulking grandam to swing away, he sent her rolling down the slope with the armpit heave and pitch that the wrestler knows makes double of his strength.

The gold-red mane on his back stood up as he nosed and mouthed the post, then he raked his flanks against it, and reared and rubbed again; ran forward a little to scan the trail, came back to rub in a new excitement, then raced like a Mad-moon buck, and came again, drove others from the post, and circled off still farther in the woods.

Then nosing a trail that to the eye said nothing, he followed it at speed. This way and that, then ever more sure, sprang through a swamp-wood thicket and into a sunny open, to see leap also from the screen a slim gray form, a Razor-back, one of his own high blood: and more, his nostrils bade him know that this was the very one that left the message on the post.

She fled, he bounded after. Across the open stretch, with Foam still nearer, a keen-eyed witness might have doubted that she ran her fastest. Who can tell? This much is sure: before the edge of woods was reached he overtook her, and she wheeled and faced, uttering little puffs, half fear, half begging for release; and face to face, a little on the slant they stood, strong Foam and slim Grizel.

Foam—A Razor-Backed Hog

There be some whose loves must slowly join their lives, who must overcome doubts and try each other long before convinced. And there be those who *know* at once when they have met the one, their only fate. This brief decree Foam gathered from the post; and Grizel was sure when gently rubbing on her cheeks she felt the ivory scimitars that are the proofs and symbols of the other mind.

She knew not what she went that day to seek, but now she knew she had found it.

THE LOVERS

The barnyard saw no more of Foam for days, for he wandered in the pleasant woods making close acquaintance with his new-found mate. The Red Squirrel on the tree limb chattered and coughed betimes as though to let them know that he was about, but they sought the farthest woods and so saw little but its shyest native folk.

Then one day as they wandered a strange noise came from the swamp. Foam moved toward the place, with Grizel, hip near, following. The way was down the hill toward a black muck swale. Coming close they found the usual belt of tall ferns. Foam pushed through these and in a moment found himself face to face with his foe, the huge black Kogar's Bear.

Foam—A Razor-Backed Hog

Foam's mane stood up, his eyes flashed with green fire, his jaws went "chop, chop" with deep, portentous sound. The Bear rose up and growled. He should have felt ridiculous, for he was coated with mud from his neck to the tip of his tail, black, sticky, smelly mud, the muddiest of mud. He must have wallowed there for hours. Yes, the Red Squirrel could have told you for hours on many different days. He was taking the cure that the wild beast takes: the second course, the one that follows the purge.

But Foam thought not of that. Here was the thing he hated and one time feared, but now feared less and less. Still he was not minded to risk a fight—not yet. The Bear, too, remembered the day of his mangled paw and the gaping wounds in his side, given by a lesser foe than this, and sullenly with growl or grunt, each slowly backed, and went his divers way.

THE WILDCAT

You see That turkey-buzzard a mile up yonder? He seems a speck to you, you poor blind human thing, but he has eyes, he can watch you as he swings, he can see your face and the way you are looking, and also he can see the Deer on the mountain miles away.

Foam—A Razor-Backed Hog

He cannot see the forest floor, for the leafy roof is over. But there are gaps in the roof, and they often give a peep of things going on below. So the Turkey-buzzard one day watched a scene that no man could have seen.

A gray-brown furry creature with a short and restless tail came gliding down a little forest trail that was the daily path of many creatures seeking to drink at the river, but Gray-coat ran each log that lay near his line of travel, then stopping at an upright limb that sprang from the great pine trunk which made his present highway, he halted in his slinking pose, rose to the full height of his four long legs, raised high his striped head, spread his soft velvety throat, white with telling spots of black, rubbed his whiskers on the high branch, rubbed his back, and gazed up into the blue sky, displaying the cruel, splendid face of a mountain Wildcat.

In three great airy wheels the Vulture swung down, down, watching still the picture through the peephole of the roof.

The Wildcat scratched his chin, then his left cheek, then his right, and was beginning all over again when a medley of sounds of voices and of many feet was heard afar, and Gray-coat's eager, alert, listening poise was a thing of power, restraint, and of wondrous grace.

Foam—A Razor-Backed Hog

The Buzzard, swinging lower, heard them, too.

The sounds came nearer; Old Gray-coat of the cruel face sprang lightly from the fallen pine to the stump where once it grew; there with the wonderful art of the beast of prey he melted himself into the stump—became nothing but a bump of bark.

The sounds still grew. Plainly a host of creatures were coming down the game trail. The Wildcat gazed intently from his high lookout. The lesser cover moved, then out there stepped a mother Razor-back with a brood of jostling, rustling, grunting, playful little Razor-backs behind her. Straying this way and that, then bounding to overtake mother, they made a little mob of roysterers; and sometimes they kept the trail, but sometimes wandered. Stringing along they came, and the bobtailed Tiger on the stump gazed still and tense, with teeth and claws all set, for here was a luscious meal in easy reach. The mother passed the stump with its evil-eyed watchman, and also the first and second of the rollicking crew. Then there was a gap in the little procession, and the Tiger gathered himself for a spring, but other sounds of feet and gruntings told that more were coming, and they rollicked after mother; another gap, and last and least of all, the runtie of the brood.

Everything was playing the Tiger's game. He

sprang. In a moment he had the little pig by the neck. Its scream of pain sent a thrill through all the band. The mother wheeled and charged. But the big cat was wise. He had made a plan. In one great scrambling bound he was high and safe on the pine stump, with the little pig squealing beneath his paws where he held it tight and remorselessly as he gazed down in cruel scorn on the tormented mother vainly ramping at the stump.

At her highest stretch she could barely touch its top edge. Beyond that was past her reach, and the big cat on the stump struck many a cruel blow with his armed paw on the frantic mother's face. There seemed no way, no hope for Runtie. But there was, and it came not from the head of the procession, as the cat had feared, but from the tail.

The Turkey-buzzard, lower yet, not only saw and heard, but even got some of the sense of shock the great cat got when the bush tops jerked and swayed and parted, and out below there rushed a huge Wild Boar.

If Cruel-face had been at all cowed by the raging mother, he would have been terror stricken now, and when that mighty beast rose up and reared against the stump, his jaws with their sabres could sweep halfway over the top, and the gray-coated villain had to move quickly to the other side, and

Foam—A Razor-Backed Hog

ever change as the Boar rushed around, but he never lost hold on the baby pig, whose squeals were getting very feeble now.

Then the silent Turkey-buzzard and the noisy applauding Red Squirrel saw a strange thing happen: The stump was beyond reach of the Boar at his highest stretch, but the great pine log was there, and three leaps away was a thick side limb that made a place of easy ascent. 'Twas here the mother scrambled up, then along the log, and now with a little leap she was on the stump and confronting the Tiger.

He faced her with a horrible snarl, a countenance of devilish rage; to scare her was his intent. What, scare a mother Razor-back, whose young is screaming "Mother, Mother, help me!" She went at him like a fury. The stinging blow of his huge paw was nothing to the lunge, slash, and heave she launched with all her vim, and the Tiger tumbled from the stump with a howl of hate, and landed on the ground, and leaped and might have escaped, but the biggest of the brood, its warrior blood stirred up by all this war, seized his broad paw and held him just a moment—just enough, for now the Boar was there.

Oh, horrors! what a shock it is, even when the fallen foe is one we hate! The mighty rush of the

Foam—A Razor-Backed Hog

Boar, the click of weapons, the hideous rumbled hate, the animal heaving sounds, the screech and chop, the flying mist of hair, the maze of swift and desperate act, the drop to almost calm, then the slash, slash, slash with sounds of rending pelt and breaking bone, and tossing of a limp form here and there, or the holding of it with both forefeet while it is mangled yet again.

The Boar grew calm, his battle madness went, and the little pigs came, one by one, to sniff and snort and run away. They had added another that day to their catalogue of smells.

And Runtie, he was lying deep in the brush on the other side of the stump. His mother came and nosed him over and nudged him gently and walked away and came again to nudge. But the brothers were lively and thirsty: she must go on with them. She raged against the fierce brute that had killed her little one. She lingered about, then led the others to the brook. Then they all came back. The little ones were once more merry and riotous. The mother came to nudge and coax the limp and bloody form, but its eyes had glazed. The father tossed the furry trash aside, and then all passed on.

These things the Turkey-buzzard saw, and I would I had his eyes, for this was a chapter in the story of Foam and Grizel that was told only by the silent little signs that it takes a hunter's eyes to see and read.

THE PORK-EATING BEAR

Why does pork-eating become so often a mania? Why does it commonly end in dire disease? We do not know. We have never heard of such penalties with any animal foods but pork. Surely the fathers of the church were wise who ruled that their people touch it not at all.

The Kogar's Bear was a pork-eater now. His range was all the valley where there were pigs, and his nightly resort was some pig-pen where the fat and tender young porkers were an easy prey, far, far better to the taste and much safer to get than the bristle-clad young rooters of the Razor-back breed. He seemed to know just when and where to go to avoid trouble and find sucklings. Of course he did not really know, but each time he raided some pig-pen the uproar of hounds and hunters for a day or more after induced him to seek other pastures, and when he happened on them his nose was sure to guide him to the pen of fatling pigs. Traps were set for him, but avoided, because he never went twice to the same pen. So the combination of shyness and keen smelling looked like profound sagacity, yet we must not scoff at it, for it gave re-

Foam—A Razor-Backed Hog

suits that seemed, and were, in a sense, the very same.

Is it not a curious fact that those who give up to a craze for some special meat always learn to prefer it a little "high," and "higher," and finally are not well pleased unless the food is positively tainted —a mass of vile corruption? And this they learn from the old-time animal habit of burying food when they have more than they need at once.

Thus it was that Scab-face, striding dark and silent through the woods by the branch, led by a smell he loved came on the unburied body of Runtie. The mother was away perforce with her living charge.

The Turkey-buzzard had not touched it, for it was fallen under brushwood. The orange and black sexton beetles were not there; it had not yet come in their department. It was a windfall for the Bear.

Reaching his long scabby nose into the thicket, he pulled it out, carried it a little way, then digging a hole he buried it deep to ripen for some future feast.

Wild animals usually remember their "cache," as the hunters call it, and come to the place when they chance in the neighborhood to see if it is all right. Thus Kogar's called next day.

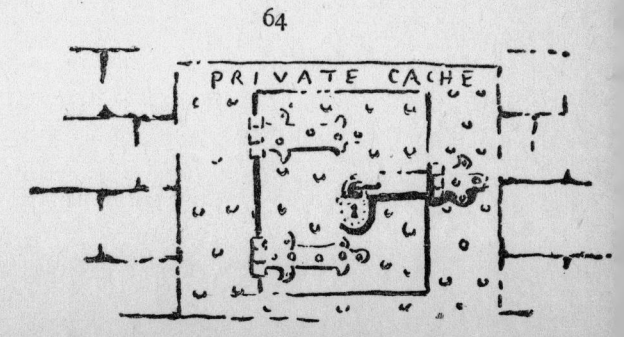

Foam—A Razor-Backed Hog

When a wild animal loses near and dear ones at a given place it goes to that place afterward for days to "mourn," as the Indians say. That is, if they are passing near, they turn aside to sniff about the place, and utter deep moans or paw up the ground, or rub the trees for a few moments, then pass on. The mourning is loudest the earliest days, and is usually ended by the first shower of rain, which robs the place of all reminiscent smells.

One day had gone since Runtie's end, and Grizel, passing on the trail, came now to mourn. And thus they met.

When a Razor-back is much afraid it gives the far-reaching tribal call for help. When it is not afraid it gives the short choppy warcry and closes with the enemy; and this is where Grizel made a sad mistake. She gave the warcry and closed. The Bear backed and dodged. They circled and sparred. The Bear would have gladly called a halt, though he was far bigger and stronger, but Grizel was bolstered up by the smell memories of the place. Her mother love was her inner strength, and still she closed; the Bear still backed till they neared the open space that lies along the high cut bank over the stream. Now was Grizel's chance, with open level ground; she charged. The

Foam—A Razor-Backed Hog

Bear sprang aside and struck with his armed paw. Had the blow landed on her ribs it might have ended her power, but it was received on her solid shoulder mass. It sent her staggering back, and as she went she gave the loud shrilling call for help, the call she should have given at first, the blast that stirs the blood of the Razor-back who hears it as the coast patrol is stirred by the cry for help. And again she fronted the Bear. Slowly turning this way and that, they faced each other, each watching for a chance. Grizel made a feint, the foe swung back, she charged. The Bear recoiled a little, braced, then swung and dodged, then as she passed he struck a mighty blow that hurled her, badly bruised and struggling, down the slope three leaps away, and over the cut bank, to splash into the stream below.

She could swim quite well, but loved it not. She splashed as she struck out, and gave no cry, for the blow had robbed her of her wind. Then the kindly stream bore her quickly down to a far and easy landing.

A moving in the bushes, a large animal sound, and on the bank there loomed a bulk of reddish black. Grizel now scrambled out and with the low short sounds of recognition they came together. But Foam had come a little late. The Bear was gone, and gone with a new-found sense of triumph, Scab-face had vanquished a full-grown Razor-back.

HILL BILLY BOGUE

Jack Prunty was raging. He walked around his new garden that morning using language that is never heard nowadays except perhaps on the golf links, certainly not permissible elsewhere. Here were lines of lettuce gone and whole patches of beets and watermelons. The asparagus bed, though not in active service, was trampled, while the cabbage patch was simply ruined.

His negro help was careful to point out that all the damage was by "hawgs"—this to prevent any suspicion lighting on the innocent. But it was not necessary. The broken fence, the myriad hoofmarks and bites taken out of turnips and cabbage were proof enough; no blame could rest on the negro or his kin.

Jack Henty was raging. He walked around his ample barnyards that morning uttering Virginianisms, as his faithful negro foreman pointed out (to prevent mistakes) that the Bear had gone here and there, and here had carried off the thoroughbred pedigreed imported Berkshire, hope of its race; and it wasn't the first they had lost, for Henty and his friends had other pens,

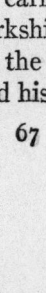

Foam—A Razor-Backed Hog

and in many raids their losses had been heavy. But this was the climax. The sow on which his hopes were built was the victim selected by the Bear

This is why Hill Billy Bogue received two invitations in one day to come with his "houn' dawgs" and win immortal fame as the defender of gardens and pens.

There were reasons for favoring Prunty. Henty was little loved: he was too rich and grasping, and had used harsh language toward Bogue, with threats of law for crimes that certainly had been committed by some one near.

So Hill Billy appeared at the Prunty home with five gaunt dogs and a new sense of social uplift. Much as the undertaker dominates all the household at a funeral, so Hill Billy at once assumed the air and authority of a commander and expert.

"Ho, ho! Wall I be goshed! Look at them for tracks—a hull family. Gee whiskers! what an ole socker! I bet yeh that was a fo'-hunder-pound Boar."

"Oh, daddy," cried Lizette, "do you suppose it was Foam?"

"Don't care if it was," said Prunty. "We can't stand this destruction; it's a case of stop right now."

Foam—A Razor-Backed Hog

The hunter kept on his examination of the trail. He was a shiftless old vagabond, useless for steady work, and a devotee of the demijohn, but he certainly knew his business as a tracker. He announced, "Just a regulation ole Razor-back family, a long-legged sow, a hatchin' o' grunters, and a Boar as big as a chicken house."

The fence was little more than a moral effect. Conscientious cows and incompetent ducks it might keep out, but to a Razor-back it was practically an invitation to attempt and enjoy. Some such thought was in Lizette's mind when she said, "Daddy, why can't we make a real fence, and a strong one that no pig could break through? It would be easy around three acres."

"Who'll pay for it?" said Prunty. "An' what's the use of a Razor-back anyway? They're no good."

"Wall," said the great man who was now combining Napoleon, Nimrod, and Sherlock Holmes, "didn't ye hear about the three little kids at Coe's school struck by a rattler and all died this week, the hull three of 'em? Rattlers is getting mighty thick up thet-a-way. Folks says it's all cause they cleaned out the Razor-backs, and I guess that's the answer all right."

Then Napoleon Nimrod Holmes Bogue began

Foam—A Razor-Backed Hog

to run the hoofmarks through the woods. The wanderings of the band had ceased. All here had followed the leader, so it was easy to keep the trail for a quarter of a mile. Hill Billy kept it; then, sure of the main fact, he went back, unchained the five gaunt hounds, worshipped in libation to his god, took rifle in hand, and swung away with the long, free stride of the woodsman.

Prunty was to head direct for Kogar's Hill, and then guided by the sounds in the valley below make for the spot where the clamor of the dogs announced at length that the Razor-backs were at bay.

Lizette went with her father.

THE HOG WARRIOR AND THE HOUNDS

The hounds showed little interest for a while, for the trail was cold, but Hill Billy kept them to it for a mile or two. Then there were plenty of signs of a pig band's recent visit, and Billy was relieved of the labor of trailing, for now the scent was fresh, and the hounds grew keen.

Then loud musical baying rang in the forest as they trailed and blew their hunting blasts. There were sounds of going in the distance, of rushing through grass and thickets, and short squeals, and some deeper sounds more guttural, and ever the baying of the hounds.

Foam—A Razor-Backed Hog

The chase swung far away, and Billy had much ado to follow. Then the sounds were all at one place, and Billy knew that the climax was at hand, the moment of all that the hunter loves, when the fighting quarry is at bay, and ready for a finish fight.

The baying of the hounds was changed as he hurried near; now it was a note of fear in some; then there was an unmistakable yell of pain, and again the defiant baying that means they are facing a quarry that they hold in deep respect.

Forcing his way through the thick brushwood, Billy got within twenty yards of the racket, but still saw nothing.

"Yap, yap, yap, yip, yip, yow, yow," went the different dogs. Then sounded the deep-chested "Gruff, gruff" of a huger animal, and a wee, small sound, a "click, click." Oh, how little it seemed, but how much it meant—the click of a Razorback's tusks—the warning that comes from a fighting Boar. The baying moved here and there, then the bushes swayed, there was a sound of rushing, there were hound yells of pain and fear, and a yelping that went wandering away to the left, and another unseen rush with a deep-toned "*Howrrr*," and nothing to be seen. It was maddening, his dogs being killed, and he could take no part.

Foam—A Razor-Backed Hog

Bogue rushed recklessly forward. In a moment he was facing a scene that stirred him. He saw the huge hog warrior charge, he saw the flashing scimitars of golden white, he saw but two dogs left—then only one, the mongrel of the pack, and the Razor-back, sighting his deadliest foe, dashed past the dog and charged. Up went the rifle, but there was no chance to aim; the ball lodged harmlessly in the mud.

Now Billy sprang aside, but the Boar was near, was swifter, stronger, less hindered by the brush. The hunter's days would have ended right there but for the remaining dog, who seized the Razor-back by the hock, and held on as for dear life.

Hill Billy saw his chance. Plunging out of the dangerous thicket to the nearest tree, he swung himself up to a place of safety, as the Boar, having slashed this wastrel of the pack, came bristling, snorting and savage, to ramp against the harboring tree, and speak his hatred of the foe in raucous, deep-breathed, grating animal terms.

LIZETTE AND AN OLD FRIEND

What joy it is to be in a high place and see the great leafy world at our feet below. What joy on a hunt, to hear the stirring hunting cry; to know that some great beast is there, and now we

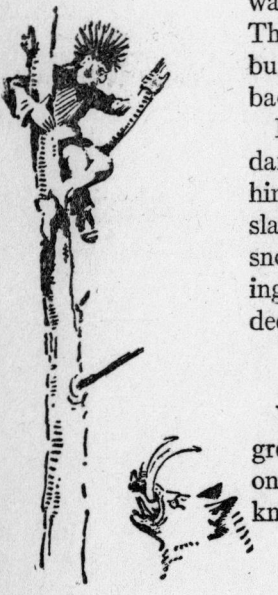

Foam—A Razor-Backed Hog

may try our mettle if we will. Some memory of his youth came back on him as Prunty with Lizette held eager harkening to the chase. How clear and close it sounded, and when the baying centred at one spot Old Prunty was like a boy, and, rushing as he should not at his age, he stumbled, slid, and fell, giving himself a heavy shock, and hurting his ankle so badly that he sat down on a log and railed in local language at his luck.

The baying of the hounds kept on. He tried to walk, then realizing his helplessness, he exclaimed: "Here, Lizette, you hurry down to Bogue and tell him to hold back for me as long as he can. I'll follow slowly. You better carry the gun."

So Lizette set off alone, guided only by the clamor of the hounds. For twenty minutes it was her sufficient guide, then it seemed to die away. Then there were a few yelps and silence. Still she kept on, and, hearing nothing, she gave a long shout that Bogue up the tree did not hear; so she tried another means, her whistle, and judging that the other hunter was coming to his rescue, Bogue shouted many things that she could not understand.

Then, seeking guidance from his voice, and offering guidance to her father, she whistled again and again. It reached them both, but it also

Foam—A Razor-Backed Hog

reached another. The great Boar raised his head. He ceased to ramp and growl. He gave an inquiring grunt. Then came anew the encouraging whistle.

From his high, wretched perch Bogue saw Lizette suddenly appear alone but carrying the rifle, and mount a log to get a view. He shouted out:

"Look out! He's going your way! Get up as high as you can and aim straight!"

It was all so plain to him, he did not understand why she should be in doubt. But she gave another loud whistle. A great red-maned form came quickly through the bushes, uttering a very familiar soft grunt. At first she was startled, then it became clear.

"Foam, Foam, Old Foamy!" she cried, and as the huge brute came trotting, his bristling crest sank down. He reared upon the log. He whispered hog-talk in his chest, he rubbed his cheek on her foot, he moved his shoulder hard against the log, and then held up his mighty hoofs arow for the pleasant rub that one time meant "French polish." Nor did he rest content till their ancient pact was carried out, and Lizette had scratched his broad and brawny back. Sitting on the log beside him, she scratched while Bogue in the tree screamed

Foam—A Razor-Backed Hog

warnings and urgings to, "Shoot, shoot, or he'll kill you!"

"Shoot, you fool!" she snorted. "I'd as soon think of shooting my big brother; and Foam wouldn't harm me any more than he would his little sister."

So the wild beast was tamed by the ancient magic, and presently the big Boar, grunting comfortably, went to his woods and was seen that day no more.

THE BEAR CLAIMS ANOTHER VICTIM

Yes, the Bear came back on a later day to his cache beside the river and the scene of his victory; there robbed the Vultures and had his horrid feast. He lingered in the neighborhood, and thus it was that fortune played for him. When next the Razor-back crew came rooting and straggling through the woods, the mother ahead and father too far astern to be a menace, they came to the fording place of the river. The little ones loved it not, held back, but mother pushed on ahead, had almost to swim in the middle. The family lingered on the bank with apprehensive grunts. One by one they screwed up their courage for the plunge, till only one was left. Finding himself alone, he set up a very wail of distress.

Foam—A Razor-Backed Hog

It reached other ears. Old Kogar's knew the cry of a lost porker. The voice was so small that his own valor was big. He glided swiftly that way. The mother pig, minded to teach her youngster a lesson of prompt obedience, paid no heed to his cry, but went on.

The left-behind one squealed still louder. The bank above his head crumbled a little under a heavy tread. There was the thud of a mighty blow, and the little pig was stilled. Then the long head and neck of Kogar's reached down and picked him from the mud. Swiftly passing up the bank, following up the slope of a leaning tree, he landed on a high ledge, and so passed over the hill.

On the other side, safer than he knew even, he sat to mouth and maul the victim, and to think in his own unthinking way, "Sweet indeed is woodland pork. The creatures are not so strong and dreadful as they seemed to me once. I fear them no longer. I will henceforth kill and eat."

THE DEFEAT OF HILL BILLY

When Hill Billy got home that night he found three of his hounds awaiting him, one of them badly cut up in body, the others very badly cut up

Foam—A Razor-Backed Hog

in spirit, for the interest they thenceforth took in hunting Razor-backs was a very small, cold, dying, near-dead thing. And start them fairly as he would it was aggravating to find how, soon or late, they took some side or crossing trail that ended where a Coon, perhaps, had climbed a tree, or a 'Possum sought the safe retreat of a crevice far in a rock.

Hill Billy might have gone to the shack of a rival hunter and borrowed more effective hounds, but that would have been admitting that his own were cowards and failures. His pride revolted at the thought. He was a true hunter at heart, not easily balked; he was strong and crafty, too, and quite able to run a trail if it seemed worth such an effort. So when a new message came from Prunty with a new tale of destruction and promises of wealth for successful service, he answered: "Wait till it comes a good rain, then I'll take the trail myself. I'll show ye."

And this was why the morning after the first heavy rain that memorable still hunt was organized. Only Prunty and Bogue took part. The hunter didn't want a crowd; this was a still hunt. Lizette's appeals for peace and a real fence were ignored. "You shall have his ivories for a bracelet; I'll get a gold band put on," was the bribe

Foam—A Razor-Backed Hog

her father offered, nearly as much to buy off himself as his daughter.

THE DAY OF JUDGMENT

Heavy rain wipes out all previous tracks. It makes the new track deep and strong. It stills all rustling leaves or crackling twigs. After heavy rain a good hunter needs no hound. Away they went, Hill Billy and Prunty, each taking a rifle often proved, for both were riflemen. They differed little in age, but Prunty was sore pressed to keep up with the lank, lithe hunter who strode ahead scanning every yard of ground for some telltale sign.

Down in the swamp were ancient marks now dim with rain. All they said, and said it feebly, was, "Yes, but some days back."

So the hunters coursed along the swamp edge and down the branch, then over the low hills, and on to Kogar's Creek, and Prunty, breathless, called a halt. Hill Billy kept on, and within a mile had found what he sought so hard, the trail of a band of Razor-backs. He followed but a little way, till he also found their leader's four-inch track, that made the rest look trivial.

"Yo, ho!" he shouted back to Prunty. "I've got him! Come on!" and Billy was off with no thought for anything but the track.

Foam—A Razor-Backed Hog

Prunty struggled along behind, but the pace was overhot for him. The answering shouts from Hill Billy became very faint; so, tired and wrathy, Prunty sat down on a log to rest and wait for something to turn up.

A quarter of an hour passed. He was breathed, and feeling better now, but there was no guiding sound to tell of the hunter's whereabouts. Another quarter of an hour, and Prunty left his log to seek the high lookout of Kogar's Hill. And getting there after a slow tramp, he sat again to wait.

Nearly an hour in all had gone, when down in the swale by the branch that fed the Kogar's Creek he heard mixed sounds of something moving in the low woods, and he made for the place.

After a short time he stopped to listen, and heard only the "jay, jay" of the Bluejay. Then once in the silence came the unmistakable shrilling of a pig in distress, the call for help. Once it came, and all was still.

Prunty pushed forward as quickly as he could, and as silently. He was nearing the open woods along the Kogar's Creek.

There were confused noises ahead, sounds of action rather than of voices, but sometimes there came voices, too: animal voices, voices that told of many and divers living things.

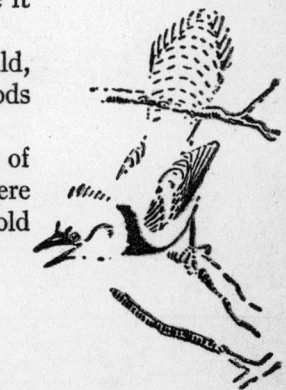

Foam—A Razor-Backed Hog

Prunty conjured up all the woodcraft of his youth. He sneaked as a Panther sneaks, lifting a foot and setting it down again only after the ground was proven safe and silent. He wet his finger to study the wind, or tossed up grass to show the breeze, and changed about so as to make an unannounced approach. He strode swiftly in the open places, and looking well to his rifle came through a final thicket where a huge down tree afforded a high and easy outlook, and mounting its level trunk he saw the setting for a thrilling scene—a face to face array of force, like hosts arrayed for battle in the olden times, awaiting but the word of onset.

There, black and fierce, was a Bear, a Bear of biggest bulk, standing half out in the open, and facing him some dozen steps away was a Boar, a Razor-back of the tallest size, but smaller than the Bear, and bearing a long scar on his face. Behind and beside the Boar was a lesser Razor-back, with the finer snout and shorter tusks of the female. Hiding in the near thicket of alder were others of their breed. At first Prunty thought but two or three, then more were seen, some very small, till it seemed a little crowd, not still, but moving and changing here and there.

Then the Bear strode in a circle toward the other

The fight with the Kogar's Bear

Foam—A Razor-Backed Hog

side of the bush, but the Boar swung round between, and the little pigs, rushing away from the fearsome brute, made many a squeak and haste to move, went quickly indeed, save one, who dragged himself like a cripple; and red streaks there were on his flank as well as a dark smear on his neck.

Thus the pair stood facing, each still and silent. Just a little curl there was on the scabby nose of the big Bear, for this was the brute of Kogar's Creek, and sometimes deep in his chest he rumbled as you hear the thunder rumble in the hills to say it will be with ye soon. And the Boar, high standing on his wide-braced legs, made bigger by the standing mane on his crested back, his snout held low, his twinkling eyes alert, his great tusks gleaming, and his jaws going "chop, chop" till the foam that gave him his baby name was flecked on the massive jowl.

The little pigs in the thicket uttered apprehensive grunts, but the big one bade his time, without a sound save the "chop" or "click" of his war gear.

There was a minute of little action, as the great ones stood, prepared, and face to face.

Who can measure the might of their moving thoughts: the Bear urged only by revenge or the lust of food, and backed by many little victories;

Foam—A Razor-Backed Hog

the Boar responding to the scream for help that stirs the fighting Boar as the fire bell stirs the fire hall horse, hastening with all the self-forgetfulness of a noble nature to help one of his kind, and finding it one of *his brood*, his very own, and, more, being harried indeed by one he held in lifelong hate? Thus every element was here supplied for a frightful clash. Power, mighty power, lust, insanity, and a doubtful courage, against lesser power with matchless courage, and the lungs and limbs of a warrior trained—Kogar's Bear and Foam of the Prunty Farm.

The big Bear moved slowly to one side, then swung in a circle around the bush, whether to make a flank attack on the Boar, or to strike at the young, mattered not; for each way the great hog swung between, resolute, head down, wasting no force in mere bluster, silent but waiting, undismayed.

Then the Bear moved to the other side, mounted a log, grunted, was minded to charge, put one paw down this side the log, and Foam charged him. The Bear sprang back. The Boar refrained. Another swing, a feint, and the Bear rushed in. Ho! Scabface, guard yourself, this is no tender youngling you've engaged.

Thud thud—thud—went the Bear's huge paws, and deep, short animal gasps of effort came. The

Foam—A Razor-Backed Hog

Boar's broad back, all bristle-clad, received the blows; they staggered but did not down him, and his white knives flashed with upward slash, the stroke that seeks the vitals where they are least ingirt with proof. The champions reeled apart. The Boar was bruised, but the Bear had half a dozen bleeding rips. Great sighs, or sobs, or heavy breathings there were from these, but from the crowded younglings just behind, a very chorus of commingled fear and wrath.

This was the first, the blooding of the fight, and now they faced and swung this way and that. Each knew or seemed to know the other's game. The Boar must keep his feet or he was lost, the Bear must throw the Boar and get a death grip with his paws ere with his hinder feet he could tear him open. The battle madness was on both.

Circling for a better chance went Kogar's, confronted still by the Boar. Again they closed, and the Bear, flinging all his bulk on Foam, would have thrown him by his weight, but the Boar was stout and rip-ripped at the soggy belly, till the Bear flinched, curled, and shrank in pain. Again and again they faced, sparring for an opening. The Bear felt safer on the log. On that he stood, and strode and feinted a charge, till Foam, impatient for the finish, forward rushed. The log was in the

Foam—A Razor-Backed Hog

way. He overleaped it, but this was not his field. The trunks that helped the Bear were baulks to him. Again they closed, and springing on his back the Bear heaved down with all his might. Slash, slash, went those long, keen, ivory knives. The Bear was gushing blood, but Foam was going down; the fight was balanced, but the balance turning for the Bear. When silent, save for the noise of rushing, another closed, another struck the Bear—Grizel was on him with her force, the slashing of her knives was quick and fast; the Bear lurched back. She seized his hinder paw and crunched and hauled; Foam heaved the monster from his back, and turned and slashed and tore. The Bear went down! Oh, Furies of the woods! What storm of fight! The silent knives or their click—the deep-voiced sob of pain and straining, the half-choked roar, the weakening struggle back, the gasp of reddened spray, the final plunge to escape, the slash, the tear, the hopeless wail—and down went Kogar's with two like very demons tearing, rending, carving. He clutched a standing tree-trunk that seemed to offer refuge. They dragged him down. They slashed his hairy sides till his ribs were grated bare. They rent his belly open, they strung his bowels out over the log like wrack weed in a storm. They knived and heaved till the dull screams died, all

Foam—A Razor-Backed Hog

movement ceased, and a bloody, muddy mass was all that was left of the Kogar's Bear.

And Prunty gazed like one who had no thought of time or space, or any consciousness but this: he was fighting that fight himself. He watched the strong hog warrior win, and felt the victory was his own. He loved him: yes, loved him as a man of strength must love a brave, hard fighter. He saw the great, big-hearted brute come quickly to himself, turn wholly calm, and the little pigs come fearfully to root and tear at the fallen foe, then rush away in fright at some half-fancied sign of life. He saw the gentleness the mates showed each to each, and ever there were little things that told of a bond of family love. Animal, physical love, if ye will, but the love that endures and fights, and still endures. And the man looked down at the thing that his hands were clutching, the long, shiny, deadly thing for murder wrought, and ready now prepared. A little sense of shame came on him, and it grew. "He saved my lil' gel, and this was my git-back." Then, again, with power returned the feelings of the day when his Lizette, the only thing he had on earth to love, came home ablaze to tell of the rattlesnake fight—with power these feelings came, and he was deeply moved as then. Her words had sudden value now. Yes, she was

Foam—A Razor-Backed Hog

right. There were other and better ways to save the crops.

His mannish joy in force and fight rose in him strong, and he blustered forth: "Gosh, what a scrap! That was the satisfyingest fight I ever seen. My! how they tore and heaved! Kill him? Gosh! you bet, for me, he can roam the swamps till he dies of a gray old age."

. . . .

The great Boar's mate turned now to lead the brood away. They rollicked off in quick forgetfulness, the wounded one came last, except that very last of all was Foam, with many rips that stood for lifelong scars, but strength unspent; and as he swung, he stopped, and glancing back, he saw his foe was still, quite still, so went.

The frond ferns closed the trail, the curtain dropped. And the Vultures swung and swung on angle wings, for here indeed was a battlefield, and a battlefield means feasting.

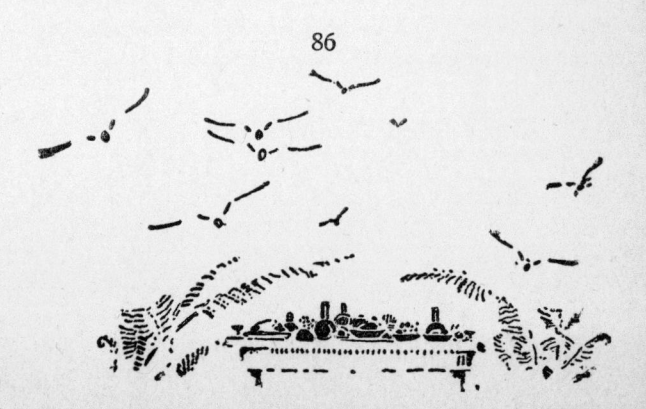

III
Way-Atcha, the Coon-Raccoon of Kilder Creek

III

Way–Atcha, the Coon–Raccoon of Kilder Creek

MOTHER Nature, All-mother, maker of the woods, that made and rejected the Bear—too big, the Deer—too obvious and too helpless in snow, the Wolf—too fierce and flesh-devouring, not deeming them the spirit of the timberland, and still essayed, till the Coon-Raccoon, the black-masked wanderer of the night and the tall timber, responsive from the workshop, came; and dowered him with the Dryad's gifts, a harmless dweller in the hollow oak, the spirit of the swamps remotest from the plow, the wandering voice that redmen know, that white men hear with superstitious dread.

Oh, help thy Singing Woodsman tell about the Coon, his kindness, his fortitude, his joy in his hollow tree, that the farmer spared because it was so hollow, and about the song he sings as he wan-

Way-Atcha, the Coon-Raccoon

ders in the night, and why he sings, and why the woodsman loves his wild and screaming yaup, even as he loves the Indian Song that holds in its bars the spirit of the burning wood.

If you will help him tell these things and make them touch the world as they have touched him, the unspeakable forester shall not work to the bitter end his sordid way, the hollow tree shall stand, and the ring-tailed hermit of the woods not pass away, nor his wind-song in the Mad Moon cease.

If he has a message, we know it not in formal phrase, but this perhaps: He is symbol of the things that certain kindly natures love; and if the nation's purblind councillors win their evil way, so his hollow tree with himself should meet its doom, it means the final conquest of the final corner of our land by the dollar and its devotees. Grant I may long be stricken down before it comes.

THE HOME-SEEKERS

March, with its ranks of crows and rolling drum calls from the woodwale, was coming in different moods to own the woods. The sun had gone, and a soft starlight on the slushy snow was bright enough for the keen eyes of the wood-prowlers. Two of them came; quickly they passed along a lying trunk, through the top of the fallen tree, across the snow to follow each convenient log as a sort of sidewalk. They were large animals—that is, larger than a Fox—of thick form, with bushy tails on which the keen night eyes of a passing Owl could see the dark bars, the tribal flag of their kind.

The leader was smaller than the other, and at times showed a querulous impatience, a disposition to nip at the big one following, and yet seemed not to seek escape. The big one came behind with patient forbearance. The singing woodsman, had he seen them, would have understood: these were mates. Obedient to the animal rule, all arrangements for the coming brood were in the mother's control. She must go forth to seek the nursing den; she must know the very time; she alone is pilot of this cruise. He is there merely to fight in case they meet some foe.

Down through the alder thicket by the stream and underbrush, and on till they reached the great stretch of timber that was left because the land was low and poor. Much of it was ancient growth, and the Coon-Raccoon—the mother soon to be—passed quickly from one great trunk to another, seeking, seeking—what?

The woodsman knows that a hollow pine is rare, a hollow maple often happens, and a hollow basswood is the rule. He might have found the har-

Way-Atcha, the Coon-Raccoon

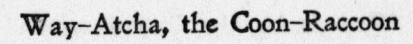

boring trunk in broad daylight, for a hollow tree has a dead top, but in the gloom the Coon seemed to go from one great column to the next with certainty, and knew without climbing them if they were not for her; and at last by the bend where the creek and river join, she climbed the huge dead maple, like one who knows.

This is the perfect lodgment of Coon-Raccoon—high up some mighty, towering tree in some deep, dangerous swamp, near running water with its magic and its foods, a large, convenient chamber, dry and lined with softest rotten wood, a tight-fit doorway, and near it some great branch which gets the sun's full blaze in day. This is the perfect home, and this was what the mother Coon had found.

THE HOME

In April the brood had come, five little ones, ring-tailed and black-masked like their parents. Their baby time was gone, and now in June they were old enough to come out on bright days, and sit in a row on the big limb that was their sunning place. Very early in life their individual characters appeared. There was the timid one whose tail was a ring too short, the fat gray one that was last to leave the nest, and the very black-masked one who was big, restless, and ready to do anything

Way-Atcha, the Coon-Raccoon

but keep quiet, the one that afterward was named Way-atcha. In their cuddling nursery days the rules of Coon life are simple. Eat, grow, keep quiet—all the rest is mother's business. But once they are old enough to leave the nest they begin to have experiences and learn the other rules.

The sunning perch was free for all, and the youngsters were allowed to climb higher in the tree among the small branches, but below the nest was a great expanse of trunk without any bark on, and quite smooth, a very difficult and dangerous place to climb, and whenever one of the youngsters made a move downward, mother ordered him back in sharp, angry tones.

Way-atcha (his mother called him "Wirrr" the same as the others, but with a little more vigor to it) had been warned back twice or thrice, but that made him more eager to try the forbidden climb. His mother was inside as he slid below the sunning limb on the rough bark and on to the smooth trunk. It was twenty times too big for his arms to grip, and down he went, clutching at anything within reach—crash, scramble, down, down, down, and splash into the deep water below.

Startled by the sudden gasp of the others, the mother hurried forth to see her eldest splashing in the brook. She hurried to the rescue, but the

Way-Atcha, the Coon-Raccoon

stream lodged him against a sandbar, he scrambled out little the worse, and made for the home tree. Mother was halfway down, but seeing him climb she returned to the row of eager faces on the branch above.

Way-atcha went up bravely till he reached the tall smooth trunk where there was no bark, and here he absolutely failed, and giving way to his despair, uttered a long, whining whimper. Mother was back at the hole, but she turned now and coming down, took Way-atcha by the neck rather roughly, placed him between her own forelegs, carried him round the smooth trunk to the side where there were two cracks that gave a claw-hold, and there boosted and kept him from falling while she spanked him all the way home.

SCHOOLING THE CHILDREN

It was two weeks later or more before mother judged it time to take them down into the big world, and then she waited for a full moon. Old Coons can do very well on a black night, but they need some light, especially at the beginning of the young one's training.

Father went down first to be ready, in case some enemy was near, and now the youngsters were taught the trick of the smooth trunk. There

Way-Atcha, the Coon-Raccoon

was only one place to climb it safely: that was where the two cracks made it possible to get clawholds well apart. Mother went first to show the way, and the youngsters followed behind.

Everything was new and surprising to them, everything had to be smelt and handled, stones, logs, grass, the ground, the mud, and, above all things, the water. The bright uncatchable water was puzzling to all except of course Way-atcha who knew, or thought he knew it, already.

The youngsters were full of glee, they chased each other along logs and tumbled each other into little holes, but mother had brought them for something more serious. They had to get their first lesson in earning a living, and this she gave them mainly by example.

Have you ever seen a Coon feeding? His way is to stand by a pool, put in both hands, groping in the mud with quick and sensitive fingers, hunting for frogs, fish, crabs, etc., while his eyes rove the woods far and near, right and left, to look for other chances or to guard against possible enemies. This was mother's way, and the youngsters looked on, more interested in the catch than in the mode.

Then they crowded up close to see better, which meant they lined up along the water's edge. It was so natural to put their hands in the water that

Way-Atcha, the Coon-Raccoon

at once they were doing as mother did. What a curious sensation to feel the mud sliding between one's fingers; then perhaps a root like a string, then a round soft root that *wriggles*. What a thrill it gives! For instinctively one knows that *that is game*, that is what we are here for. And Way-atcha, who made the find, clutched the pollywog without being told, seized it in his teeth and got *chiefly* a mouthful of mud and sand. He sputtered out everything, mud, pollywog, and all. Mother took the flopping silver-belly, gravely washed it in the clear water, and gave it back to be gobbled by Way-atcha. Now he knew. Thenceforth he dropped easily into the habit of his race, and every bite was religiously washed and cleaned before being eaten. The shy brother with the short tail was too timid to go far from mother, and what he learned was little. The other two were quarrelling over a perfectly worthless old bone. Each "found it first," and the winner had a barren victory. Grayback was far out on a log over the water, trying to claw out the reflection of the moon, but Way-atcha, intoxicated by success, was now keen to keep on hunting. Down along the muddy margin he paddled, eagerly glancing this way and that, just like mother, feeling in all the mud, straining it through his fingers, just

Way-Atcha, the Coon-Raccoon

like mother, lifting up a double handful to smell, just like mother, clutching at some worthless root that seemed to wriggle, then sputtering it out with a growl, just like father. It was fun, every bit of it, and when at length his active little fingers clutched the unmistakable smooth and wriggly body of a frog that was hiding in the mud, Way-atcha got such a thrill of joy that all the hair on his back stood up, and he gave the warwhoop of the Coon-Raccoon, which is nothing more than a growl and a snort all mixed up together. It was a moment of triumph, but Way-atcha did not forget the first lesson, and that frog was washed as clean as water could make him before the hunter had his feast.

This was intensely exciting, there was limitless joy in view, but a sudden noise from father changed it all. He had been scouting far down the river bank while the youngsters played along the creek near mother. Now he gave a signal that mother knew too well, a low puff, like "Foof," followed by a deep grunt. Mother called the youngsters with a low grunting. They knew nothing at all of what it was about, but the sense of alarm had spread instantly among them, and a minute or two later there was a regular procession of furry balls climbing the great maple, following the two cracks, right up to tumble into their comfortable bed.

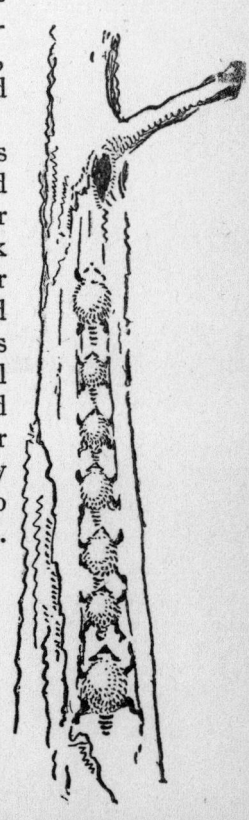

Way-Atcha, the Coon-Raccoon

Faraway down the river came a deep booming sound, the roaring of some terrible animal, no doubt. Mother listened to it from the door. Presently father came scrambling up the trunk a little wet, because he had swum the river, after laying a trail to take the enemy away, and had come home by a new road along the top of a fence, so that no trail was left and the baying of that awful hound was lost faraway in the woods.

That night Way-atcha had met and felt some of the big things that shape a Coon's life: the moonlight hunt, the vigilant mother, the fighting father, the terrible hound, the safe return home protected by a break in the trail. But he did not think about it. He remembered only the joy of clutching that fat, wriggling, juicy frog, and next night he was eager to be away on another hunt.

THE MYSTERIOUS WARNING

Many animals have a sixth sense, a something that warns them that there is danger about, a something that men once had, and called "a far sense of happenings" or a "sense of luck." This seems to be strongest in mothers when they have their young. And when the next night came Way-atcha's mother felt uneasy. There was something wrong. She delayed going down the round stair-

Way-Atcha, the Coon-Raccoon

way and lay watching and listening on the sunning branch till every one was very cross and hungry. Way-atcha was simply overcome with impatience. Father went down the trunk but soon came up again. The children whimpered, but mother refused to budge. Her quick ears were turned once or twice toward the river, but nothing of note was heard or seen. The moon had set, and at length in the darkest hours the mother led her family down the big trunk. All were hungry, and they rushed heedlessly along the bank, dabbling and splashing. Then Way-atcha caught a frog, and little Ring Short a pollywog. Then all had caught frogs, and it seemed the whole world was one big joyous hunt without a care or a worry.

Now out on a sandbar Way-atcha found a new kind of frog. It looked like two flat bones lying side by side, but the smell was pleasant. He reached out, and at once the two bones closed together on his toes, squeezing them so hard that he squalled out, "Mother, Mother!" Mother came running to help, of course, while Way-atcha danced up and down in pain and fear. But the old one had seen mussel clams before. She seized the hard thing in her teeth, crushed the hinge side, and ended the trouble. Now Way-atcha had the pleasure of picking out the meat from the sharp bits of shell, washing them clean in the river, and gobbling them as a new kind of frog, and everything seemed very well to him.

But father climbed a root and snuffed, sniffed, and listened, and mother studied all the smells and trails that were along the pathway farther from the river bank. She had had little time for hunting. Her secret sense was strong on her, and she gave the signal to return.

The youngsters followed very unwillingly. Wayatcha was almost rebellious. There seemed in his judgment to be every reason for staying and none whatever for going home. But the best of judgment must yield to superior force. Mother's paws were strong and father could be very rough. So the seven fur balls mounted the smooth maple stairway as before.

The Red Fox of the hillside yapped three times, a little song sparrow sang aloud in his dreams not far from the great maple, and the Coon mother heard without heeding. Then later came another sound, quite low and distant, feeble indeed. The young seemed not to hear it, but it set the mother's hair on end. It was a different note, coming from anywhere in the north: the harmless wind made just such noises at times, but in this were also sharp

Way-atcha with his Mother and Brothers hunting in the moonlight

cracks, like blows struck on wood, and once or twice yelps that must have been from dogs.

The sounds came nearer and louder, red stars appeared among the trees, and soon a band of men out with dogs came menacing every living prowler in the woods. The fresh Fox track down below diverted the attention of the dogs so they did not come near the Coon tree, and mother knew that they had escaped a great danger that night.

THE HUNTERS

The following evening Mother Coon looked forty ways and sniffed every breeze that blew, while the moon swung past four trees quite near the door before she would let the family go on their regular hunt. They supposed, of course, she would lead down the usual way by the creek, but she did not. She moved in a new direction upstream, nor would she stop to hunt, but pushed on. They reached a stretch of bank where frogs went jump, jump, at every bank of sedge. It seemed most promising, but mother still pushed on. Then a loud noise like rising wind was heard, only sometimes it splashed like a frog or even a muskrat. Then they came to the thing that made it, the creek itself, jumping over a rocky ledge into a pool, sparkling in the moonrays, noisy in the night. Mother held

Way-Atcha, the Coon-Raccoon

them back a little while she looked hard ahead and around. Then she crouched; her hair rose up; she growled. Father came alongside. The youngsters had no desire now to rush ahead. There around the water so full of game were other hunters, splashing, catching frogs, and feasting. They were in size like Way-atcha's people, and when the tail of one was turned there surely were the seven rings that make the tribal flag of Coon-Raccoon.

But some one was trespassing. Which family owned this hunting? That is always a serious question in the woods. Father Coon stood up very high on his legs, puffed out his hair, and walked forward from the cover, along the open margin. There was a noisy rush of the other family, then three young in it went whimpering to their mother, and their father stood up high, puffed out his hair, and came marching stiffly and openly toward Way-atcha's father. Each gave a low growl, which meant, "Here you, get out of this or I'll make you!" Then, since neither got out, they squared up face to face. Each felt that he himself was right, and the other all wrong. Each felt that he must protect his family and drive the trespassers away; and so they stood and glared at each other, while the young ones of each crowded closely behind their mothers.

This is the animal law of range. The first finder

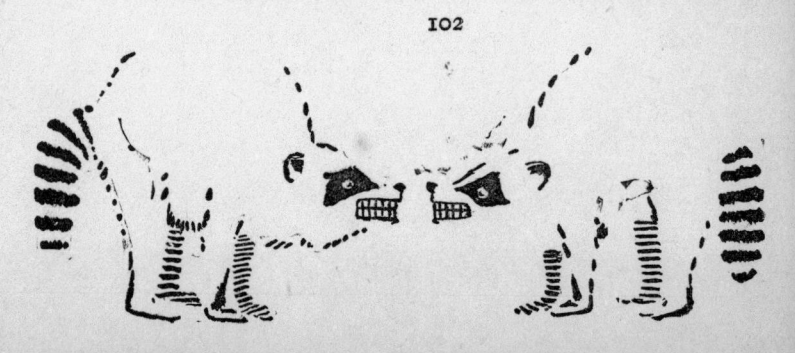

Way-Atcha, the Coon-Raccoon

owns it, if he marks it at leading points, using for this the scent glands near the tail that nature gave for just such purposes. If two hunters have equal claims, they fight, and the stronger holds it. Way-atcha's people, as it chanced, had not marked the hunting ground for weeks, so their musk marks were nearly washed away. The other family came later, but had used it much, and marked it, too. The rival claims were balanced. Nothing now but a fight could settle it.

And this is the Coon's chief mode of fight: close on the enemy, offering the well-defended neck or shoulders to his attack, seize him around the waist and throw him so he will *fall on you;* for the under Coon has the best chance to rip open his enemy's belly with hind claws, which are free; holding him with fore claws which are free, his teeth have free play at the enemy's throat, which is exposed.

So Way-atcha's black-masked sire came edging on, a little sidewise, and the Coon of the Pool having sized up the other as bigger than himself, held back a little, fearing to close at once.

Old Black Mask made a pass; the Pool Coon parried. They dodged round and round, neither gaining nor giving ground. Another pass, then Black Mask's footing slipped, the Pool Coon closed, and the fight was on. But neither got the grip he sought. Their powers were nearly even. They rolled and tugged, while their families squalled, and in a moment both went reeling, and splash, into the deep, cool pool. There is nothing like cool water for cooling. The fighters broke apart, and when they scrambled out they both felt a wonderful change. They had no more desire to fight. Each now was indifferent to the fact that the other was hunting on his grounds. They were in truth cooled off.

There were some angry looks perhaps, and a few low growls, but each with his family set about hunting round the pond, one keeping the thickwood side, the other the open side.

This was the beginning, and in time they all became good friends, for the hunting was plenty for both. The children feasted till their bodies were quite round in front and they were glad once more to climb their big smooth tree.

THE WAYWARD CHILD

Way-atcha strongly disapproved of many things his mother did. If she wished to go downstream when his plan was to go up, she must be wrong. If she was hindered by some trifling noise from going to get supper at supper time, it meant senseless annoyance for all. If she was afraid of that curious musky smell on a stone by the shore, well! he was not, and that was all about it.

They had gone for their usual supper hunt one night. After smelling the wind, mother had decided on going downstream, but Way-atcha had been enjoying visions of the pool with its varied game.

He held back, and when his mother called, he had followed only a little way. Then his keen eyes sighted a movement in the edge of the near water. He sprang on it with the vigor of a growing hunter, and dragged out a fine big crawfish. Then he proceeded to wash it thoroughly and ate it body and bones, not heeding the call of his mother as she led the others away. He was perfectly delighted with himself for this small victory, and felt so set up and independent that he turned in spite of mother's invitation and set out to visit the upper pool as he had planned.

After one or two little captures he reached the jumping water. That very day another visitor had been there. Indian Pete, a trapper, had found the pool, and all about it had seen the tracks of Coon and Muskrat. At this season fur is worthless, but Pete used these creatures for his food, so hid a big steel trap in the mud, and on a little stick farther out in the water he rubbed a rag with a mixture of animal oils and musk.

Way-Atcha, the Coon-Raccoon

Ho, ho! there it was again, that very smell that poor timid mother was in such fear of. Now he would examine it. He came down to the place, then sniffed about, yielded to his habit of feeling in the mud as he glanced this way and that, when *snap, splash,* and Way-atcha was a prisoner held firmly by one paw in a horrible trap of steel.

Now he thought of mother, and raised the long soft *whicker* that is the call of his kind, but mother was far away. He himself had made sure of that, and he remembered the clam shell, but all his efforts to pull away or bite off that horrid hard thing were useless; there it clung to his paw, and hanging to it was a sort of strong twisted root that held him there. All night long in vain he whickered, whimpered, and struggled. He was worn out and hoarse as the sun came up, and when Indian Pete came around he was surprised to find in his Muskrat trap a baby Coon, nearly dead with cold and fright, and so weak that he couldn't even bite.

The trapper took the little creature from the trap and put him alive in his pocket, not knowing exactly what he meant to do with him.

On the road home he passed by the Pigott homestead and showed his captive to the children.

The little Coon was still cold and miserable, and when put into the warm arms of the oldest girl he

Way-Atcha, the Coon-Raccoon

snuggled up so contentedly that he won her heart and she coaxed her father into buying Way-atcha, as the Indian named the captive in his own tongue.

Thus the wanderer found a new and very different home. He was so well taken care of here that in a few days he was all right again. He had children to play with instead of brothers and sisters, and many curious things to eat instead of frogs, but still he loved to dabble his own brown paws in the mud or anything wet whenever he could get the chance. He did not eat milk and bread like a cat or other well-behaved creature; he always put in his paws to fish out the bread, bit by bit, and commonly ended by spilling the milk.

A MERRY LIFE ON THE FARM

There was one member of the household that Way-atcha held in great fear; that was Roy the sheep-dog, house-dog, watch-dog, and barnyard guard in general. When first they met Roy growled and Way-atcha chirred. Both showed in the bristling shoulder hair that they were deeply moved; each in the smell of the other was instinctively aware of an enemy in an age-long war. The Pigott children had to exercise their right of eminent domain to keep the peace; but the peace was kept. Roy learned to tolerate the Coon in time, the Coon

Way-Atcha, the Coon-Raccoon

became devotedly fond of Roy, and not two weeks had gone before Way-atcha's usual napping couch was right on Roy's furry breast, deep in the wool, cuddled up with all the dog's four legs drawn close against him.

As he grew stronger he became very mischievous. He seemed half monkey, half kitten, full of fun always, delighted to be petted, and always hungry, and soon learned where to look for dainties. The children used to keep goodies in their pockets for him, and he learned that fact so well that when a stranger came to the house Way-atcha would gravely climb up his legs and seek in all his pockets for something to eat.

On one occasion he had been missing for some hours, always a suspicious fact. When Mrs. Pigott went into the storeroom, stocked now with the summer preserves, she was greeted with the whining call of Way-atcha, more busy than words can tell. There he was wallowing up to his eyes in plum jam, digging down into a crock of it like a washwoman into her tubs, feeling and groping for what? He had gorged himself till he could eat no more, and now prompted by his ancient woodland memories he was gropping with his paws among the jam and juice to capture all the plum stones, each in turn to be examined and cast aside.

Way-Atcha, the Coon-Raccoon

The floor was dotted with stones, the shelf was plastered with the jam of the many pots examined. The Coon was unrecognizable except for his bright eyes and face, but he came waddling, whining, slushing down from the shelf across the floor to climb up Mrs. Pigott's dress, assured, he believed, of a cordial welcome. Alas! what a cruel disappointment he got!

One day Mr. Pigott set a hen with thirteen eggs. The next day Way-atcha was missing. As they went about calling him by name they heard a faint reply from the hen-house, the gentle "whicker" that he usually gave in answer. On opening the door, there they saw Way-atcha sprawling on his back in the hen's nest perfectly gorged, and the remains of the thirteen eggs told that he was responsible for a piece of shocking destruction. Roy was the proper guardian of the hen-house. No tramp, no Fox, no Coon from the woods could enter that while he was on guard. But alas! for the conflict of love and duty: in his perplexity the dog had unwittingly followed the plan of a certain great man who said, "In case of doubt, be friendly."

Farmer Pigott bore with Way-atcha for long because the children were so fond of the little rascal. But the climax was reached one day when the Coon, left alone in the house, discovered the

Way-Atcha, the Coon-Raccoon

ink bottle. First he drew the cork and spilled the ink about, then he dabbled his paws in it after his usual manner, and found a new pleasure in laying the inky paws on anything that would take a good paw-mark. At first he made these marks on the table, then he found that the children's school books were just the things and gave much better results. He paw-marked them inside and out, and the incidental joy of dabbling in the wet resulted in frequent re-inking of his paws. Then the wall paper seemed to need touching up. This lead to the window curtains and the girls' dresses, and then as the bedroom door was open Way-atcha scrambled on the bed. It was just beautiful the way that snow-white coverlet took the dear little paw-marks as he galloped over it in great glee. He was several hours alone, and he used up all the ink, so that when the children came in from school it looked as though a hundred little Coons had been running all over the place and leaving black paw-marks. Poor Mrs. Pigott actually cried when she saw her beautiful bed, the pride of her heart. But she had to relent when Coonie came running to her just the same as usual, holding out his inky arms and whining *"errr err"* to be taken up and petted as though he were the best little Coon in the world.

Way-Atcha, the Coon-Raccoon

But this was too much. Even the children had no excuse to offer; their dresses were ruined. Way-atcha must go; and so it came about that Indian Pete was sent for. Way-atcha did not like the looks of this man, but he had no choice. He was bundled into a sack and taken away by the half-breed, much to Roy's bewilderment, for he disliked the half-breed and despised his dog. Why they should let *that stranger* carry off a member of *his family* was a puzzle. Roy growled a little, sniffed hard at the hunter's legs, and watched him without a tailwag as he went off with the bulging bag.

THE ANCIENT FOE

It was the end of summer now, the Hunting Moon was at hand; the hunter had a new hound to train, and here was the chance to train him on Coon. Way-atcha had no claim on Peter's affection, and nothing educates a dog for Coon so much as taking part in a Coon run and kill.

This was then to be the end of Way-atcha. The trapper would use him, sacrifice him, to train his hunting dog. As he neared his shanty that dog came bounding forth, a lumbering half-breed hound, with a noisy yap which he uttered threefold when he sniffed the sack that held Way-atcha.

Way-Atcha, the Coon-Raccoon

And this was the way of the two: in the log stable the Coon was given a box, or little kennel, where he could at least save his life from the dog. Howler was brought in on a chain and encouraged to attack the Coon with loud "sic hims." Brave as a lion, seeing so small a foe, he rushed forward, but was held back with the chain, for it was not time for a "kill." Many times he charged, to be restrained by his master.

Way-atcha was utterly puzzled. Why should those other two-legged things be so kind and this so hostile? Why should Roy be so friendly and this yellow brute so wicked and cruel? Each time the big dog charged, poor little Way-atcha felt in him the fighting spirit of his valiant race stirred up, and faced the brute snarling and showing all his teeth.

But he would quickly have been done to death by the foe had not the half-breed held the chain. Only once was the dog allowed to close. He seized the Coon cub by the neck to give the death shake, but nature gave the Coon a strong, loose skin. The shake was scarcely felt, and Way-atcha clamped his teeth on Howler's leg with a grip that made him yell; then the half-breed dragged the dog away. That was enough for lesson No. 1. Now they hated each other; the bitter feud was on.

Way-Atcha, the Coon-Raccoon

Next day a lesson was given again for both, and both learned other things: Way-atcha that that hole, the kennel, was a safe refuge; the cur, that the Coon could clutch as well as bite.

The third day came and the third lesson. Waiting for the cool of the evening, the hunter dropped the Coon into a bag, took down his gun, called the noisy dog, and made for the nearest stretch of woods, for the trailing and treeing of the Coon was to be the climax of the course of training.

Arrived at the timberland, Pete's first care was to tie the dog to a tree. Why? Certainly not out of consideration for the Coon, but for this: the Coon must be allowed to run and get out of sight, otherwise the dog does not try to follow it by track. Once he has to do this to find his prey, his own instinctive prompting makes him a trailer and he follows till he sights the quarry, then attacks, or if it trees, as is usual, he must ramp and rage against the trunk to let the hunter know the Coon is there. This is the training of a Coon dog; this was the plan of Indian Pete.

So the dog was chained to a sapling; the Coon was carried out of reach, and tumbled from the sack. Bewildered at first, but brave, he glared about, then seeing his tall enemy quite near he rushed open-mouthed at him. The half-breed ran away in some alarm, but laughing. The dog rushed at the Coon till the chain brought him up with a jerk, and now the Coon was free from all attack, was free to run. And then how he ran! With the quick instinct of a hunted race, he dashed away behind a tree to get out of sight, and, zigzagging, bounded off, seeking the thickest cover, running as he never had run before.

Back came the half-breed to release the dog. Tight as a guy-rope was the chain that held that crazy, raging cur, so tight the chain that he could not get the little slack he needed to unhook the snap. Cursing the dog, jerking him back again and again, he fumbled to unhook the snap; and as he jerked and shouted, the dog jerked more and barked, so made it harder. Two or three minutes indeed he struggled to release the chain, and then he had to catch and hold the dog so as to free him by slipping his collar. Away went the dog to the place where last he saw the Coon.

But the victim was gone; those precious three minutes meant so much, and responsive to the hunter's "sic him" "sic him" the dog raced around. His nostrils found the trail, instinctively he yelped, then followed it, at every bound a yelp. Then he lost it, came back, found it again, and yelped, and slowly followed, or if he went too fast he lost

Way-Atcha, the Coon-Raccoon

it. And Pete ran, too, shouting encouragement, for all of this was in the plan. The Coon no doubt was running off, but soon the dog would find him, and then—oh, it never fails—the Coon climbs up the easiest tree, which means a small one always; the dog by yapping down below would guide the man, who coming up would shoot the Coon, which falling disabled would be worried by the dog, who thus has learned his part for future cooning, and thenceforth flushed with victory be even keener than his master for the chase.

Yes, that was the plan; it had often worked before, and did so now, but for one mishap. Way-atcha did not climb a slender tree. As soon as he was far away, thanks to that fumbled chain, and heard the raging of the two behind, he climbed the sort of tree that in his memory had been most a thing of safety to him. The big hollow maple was the haven of his youth, and up the biggest tree in all the woods he clambered now.

His foes came on; the dog was learning fast, was sticking to the trail. His master followed till they reached the mighty sycamore, and "Here," said Howler, "we have treed him!" What the half-breed said we need not hear. He had brought his rifle, yes, but no axe. The Coon was safe in some great cavernous limb, for nowhere could they see him, and the tree could not be climbed by man. The night came down and Pete with his yapping dog went home defeated.

THE BLESSED HOLLOW TREE

So luck was with Way-atcha, luck and the influence of his early days, that built in with his nature the secret of his race: this is their true abiding place—the hollow tree. The slender second growth most often near is a temptation and a snare, but the huge hollow trunk is a strong fortress and a sure salvation.

Rested and keen was he, when the blackest hours came with a blessed silence; so forth he went and after many a "hark" and "spy" he swung himself to the ground in the big woods and galloped away and away, nor stopped to feed till he found himself far in the wide swamplands of Kilder Creek, in the home of his early days and the land of his kindred.

A Coon coming back after months away is a stranger to his people. His form is forgotten or changed, his place is filled. Only one thing holds among this folk of smells, that is his smell, that was his passport, the proof that he was theirs, and slowly he "came back," not as the young of such a one, but as a tribal member in good standing, and

Way-Atcha, the Coon-Raccoon

with them ever learning, and teaching too, till the inner urge asserts itself and he breaks with the band, to cleave to a mate from the band. So they leave their kind, and seek, as their parents sought, some quiet spot where huge and hollow trunks hold yet the ground, where the precious land is made beautiful by its very worthlessness. And here, by the All-mother led, they raise their brood and teach a little more than they were taught, for times have changed. The leagues of big tall woods are gone, only the skimpy remnants by the water stay, only the useless trunks on the useless land, as ploughmen think. They give no harbor to the one-time forest kings, but lure the black-masked dweller of the hollow trunk, and wise is he with growing need for wisdom. He comes not forth by day; he goes not far by night. He runs the top of every fence, so leaves a broken trail. He lives on woodland creekside food. He shuns all clash with men. He never shows himself to them unless they chance to know his way. High in the noonday sun he lies at times to take the sunning that is balm for many an ill; and in the night, when the moon is sinking, he may splash and forage by the swampy shores. There tracks of divers size next day give record of the night prowl. But ye may not see him unless by rare mischance; he is more alert than you, and ready to vanish in his hollow tree, for the world has many hunting dogs, with but one Roy. He knows you not, but he knows that there is many an Indian Pete.

Ye long to meet and know him, oh, ye Kindly Singing Woodsmen! Ye guarantee respect, yea, reverence, for the Dryad of the hollow trees! Would I might be your introducing guide!

I have sought, sought lovingly, to meet him in the low, wet woods of Kilder Creek. Many times have I put tempting corn in forks and other altars as my offering to the Ringtail. And the corn is always gone, I never know just how, but I see at divers times and trails the marks of that dexterous human-fingered paw, or the mussel shell with broken hinge, or the catfish fins, and know that still he dwells close by, that still he scoffs at bellowing hounds, nor has deep fear of any but the shameless axe that would steal his consecrated tree. What would I not give to have him let me see him as one sees a nearby Friend; but that is what he will not. All my privilege is this: to see the pattered pigmy human tracks when in the hours of morning sun I seek along the lake, or sometimes, when the autumn's night is black, I get the long-drawn rolling song, "*Whill-ill-ill-a-loo, whill-illl-ill-a-loo, whill-*

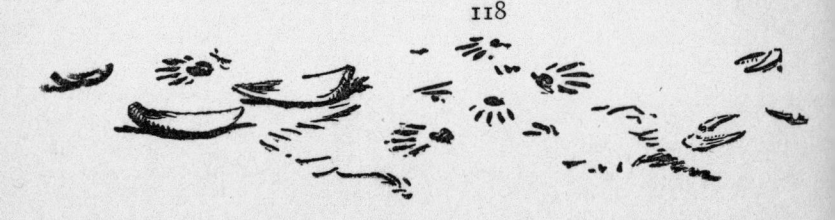

Way-Atcha, the Coon-Raccoon

a-loo," the love song of Way-atcha the Ringtail Coon-Racoon that wanders still, makes love and lives, like the remaining prophet of a bygone simple faith, that being true, will some time come again to rule, but is waiting, hiding, waiting now, till the fire has passed away.

IV
Billy, the Dog That Made Good

IV
Billy, the Dog That Made Good

SILLY BILLY

HE WAS the biggest fool pup I ever saw, chuck full of life and spirits, always going at racing speed, generally into mischief; breaking his neck nearly over some small matter; breaking his heart if his master did not notice him, chewing up clothing, hats, and boots, digging up garden stuff that he could not eat, mistaking every chicken in the barnyard for legitimate wild game, going direct from mud in the road to frolic in the parlor, getting kicked in the ribs by horses and tossed by cows, but still the same hilarious, rollicking, endlessly good-natured, energetic fool pup, and given by common consent the fit and lasting name of "Silly Billy."

It was maddening to find on the first cold morning that he had chewed up one's leather glove, but it was disarming to have that irrepressible, good-

Billy, the Dog That Made Good

natured little idiot come wagging his whole latter end south of the short ribs, offering the remaining glove as much as to say that "one size was enough for any one." You *had* to forgive him, and it did not matter much whether you did or not, for the children adored him. Their baby arms were round his neck as much of the time as he could spare from his more engrossing duties, and, in a figurative sense, those protecting arms were around him all the time. As their father found out, when one day the puppy pulled down a piece of sacking that hung on the smokehouse pipe, upsetting the stove and burning up the smokehouse and all the dry meat in it. Bob Yancy was furious, his whole winter's meat stock gone. He took his shotgun and went forth determined to put that fool dog forever out of mischief. But he met the unexpected. He found his victim with two baby arms about his fuzzy neck: little Ann Yancy was hugging her "doggie," and what could *he* do? "It's my Billy! You shan't touch him! Go way, you naughty Daddy!" And the matter ended in a disastrous defeat for daddy.

Every member of the family loved Silly Billy, but they wished from the bottom of their hearts that he might somehow, soon, develop at least a glimmer of common dog sense, for he was already past the time when with most bull terriers the irresponsible exuberance of puppyhood is ended. And though destined to a place among his master's hunting dogs, he, it was judged, was not yet ripe enough.

Bob Yancy was a hunter, a professional—there are a few left—and his special line was killing Bears, Mountain Lions, Lynxes, Wolves, and other such things classed as "varmints" and for whose destruction the state pays a bounty, and he was ever ready to increase the returns by "taking out" amateur hunters who paid him well for the privilege of being present.

Much of this hunting was done on the high level of "the chase." The morning rally, the far cast for a trail, the warming hunt, the hot pursuit, and the finish with a more or less thrilling fight. That was ideal. But it was seldom fully realized. The mountains were too rough. The game either ran off altogether, or, by crossing some impossible barrier, got rid of the hunters and then turned on the dogs to scatter them in flight.

That was the reason for the huge Bear traps that were hanging in Yancy's barn. Those dreadful things would not actually hold the Bear a prisoner, but when with a convenient log they were gripped on his paw, they held him back so that the hunters, even on foot, could overtake the victim.

Billy, the Dog That Made Good

The dogs, however, were the interesting part of the pursuit. Three kinds were needed: exquisite trailers whose noses could follow with sureness the oldest, coldest trail; swift runners for swift game, and intelligent fighters. The fighters had, of course, to be brave, but intelligence was more important, for the dogs are expected to nip at the bayed quarry from behind and spring back from his counter blow rather than to close at final grips.

Thus there were bloodhounds and greyhounds as well as a bulldog in the Yancy pack, and of course, as always happens in a community of diverse bloods, there were some half-castes whose personal worth had given them social prestige, and was accepted as an offset to doubtful pedigree. Most of the pack had marked personality. There was Croaker, a small lady hound with an exquisite nose and a miserable little croak for a bay. You could not hear her fifty feet away, but fortunately Big Ber was madly in love with her; he followed her everywhere and had a voice like the bell for which he was named. He always stuck close to Croaker and translated her feeble whispers into tones that all the world within a mile or two could understand.

Then there was Old Thunder, a very old, very brave dog, with a fine nose. He was a combination of all good gifts and had been through many fights,

Billy, the Dog That Made Good

escaping destruction only thanks to the admirable sagacity that tempered his battle rage. Though slow and feeble now, he was the acknowledged leader of the pack, respected by dogs and men.

THE PROFESSIONAL ROUGH

The bulldog is more conspicuous for courage than discretion, so that the post of "bulldog to the pack" was often open. The last bulldog had been buried with the bones of their last Grizzly. But Yancy had secured a new one, a wonder. He was the final, finished, and perfect product of a long line of fighting bulldogs kept by a famous breeder in another state. And when the new incumbent of the office arrived it was a large event to all the hunters. He was no disappointment: broad of head and chest, massive in the upper arm and hard in the flank, a little undershot perhaps, but a perfect beast of the largest size. Surly and savage beyond his kind, the hunters at Yancy's knew at once that they had a fighting treasure in the Terrible Turk.

It was with some misgiving that he was turned loose on the ranch. He was so unpleasant in his manner. There was a distinct lack of dogginess about him in the gentle sense, and never did one of his race display a greater arrogance. He made no pretence of hiding his sense of contemptuous superiority, and the pack seemed to accept him at his own value. Clearly they were afraid of him. He was given the right of way, avoided indeed by his future comrades. Only Silly Billy went bounding in hilarious friendliness to meet the great one; and a moment later flew howling with pain to hide and whimper in the arms of his little mistress. Of course, in a world of brawn, the hunters had to accept this from their prizefighter, and see in it a promise of mighty deeds to come in his own domain.

In the two weeks that passed about the ranch the Terrible Turk had quarrelled with nearly every hound in the pack. There was only one indeed that he had not actually injured: that was Old Thunder. Once or twice they confronted each other, as when Thunder was gnawing a bone that the Turk seemed to want, but each time Thunder stood his ground and showed his teeth. There was a certain dignity about Thunder that even a dog will feel, and in this case, without any actual conflict, the Terrible Turk retired, and the onlookers hoped that this argued for a kindly spirit they had not hitherto seen in him.

October was glowing on the hills, and long unwonted peeps of distant snowpeaks were showing themselves through thinning treetops when word came that Old Reelfoot, a famous cattle-killing

Billy, the Dog That Made Good

Grizzly, had reappeared in the Arrow-bell Cattle Range, and was up to his old tricks, destroying live stock in a perfect mania for destruction. There was a big reward offered for the destruction of Reelfoot, several times that held out for an ordinary Bear. Besides, there was really a measure of glory attached to it, for every hunter in the country for several years back had tried to run Reelfoot down, and tried in vain.

Bob Yancy was ablaze with hunter's fire when he heard the news. His only dread was that some rival might forestall him. It was a spirited procession that left the Yancy Claim that morning, headed for the Arrow-bell Ranch; the motley pack straggling along or forging ahead till ordered back in line by the huntsman. There was the venerable Thunder staidly trotting by the heels of his old friend Midnight, Yancy's coal-black mare; and just before was the Terrible Turk with his red-rimmed eyes upturned at times to measure his nearness to the powerful black mare's hoofs. Big Ben was fast by Croaker, of course, and the usual social lines of the pack were all well drawn. Next was a packhorse laden with a huge steel Bear trap on each side, then followed packhorses with the camping outfit and other hunters, the cook, and the writer of this story.

Billy, the Dog That Made Good

Everything was in fine shape for the hunt. Everything was fitly ordered and we were well away when a disconcerting element was tumbled in among us. With many a yap of glee, there, bounding, came that fool bull terrier, Silly Billy. Like a June-bug among honeybees, like a crazy schoolboy in a council room, he rollicked and yapped, eager to be first, to be last, to take liberties with Thunder, to chase the Rabbits, to bay the Squirrels, ready for anything but what was wanted of him: to stay home and mind his own business.

Bob might yell "Go home!" till he was hoarse. Silly Billy would only go off a little way and look hurt, then make up his mind that the boss was "only fooling" and didn't mean a word of it, and start in louder than ever. He steered clear of the Turk but otherwise occupied a place in all parts of the procession practically all the time.

No one wished him to come, no one was willing to carry him back, there was no way of stopping him that little Ann would have sanctioned, so Silly Billy came, self-appointed, to a place on the first Bear hunt of the season.

That afternoon they arrived at the Arrow-bell Ranch and the expert Bear-man was shown the latest kill, a fine heifer barely touched. The

Billy, the Dog That Made Good

Grizzly would surely come back for his next meal. Yes, an ordinary Grizzly would, but Reelfoot was an extraordinary animal. Just because it was the Bear fashion to come again soon, he might not return for a week. Yancy set a huge trap by this "kill" but he also sought out the kill of a week gone by, five miles away, and set by that another gaping pair of grinning cast steel jaws.

Then all retired to the hospitable ranch house, where Turk succeeded in mangling a light-weight sheep-dog and Silly Billy had to be rescued from a milky drowning in the churn.

Who that knows the Grizzly will be surprised to hear that that night brought the hunters nothing, and the next was blank? But the third morning showed that the huge brute had come in craftiness to his older kill.

I shall not forget the thrills of the time. We had passed the recent carcass near the ranch. It lay untouched and little changed. We rode on the five miles to the next. And before we were near we felt there was something doing, the dogs seemed pricked up, there was some sensation in the air. I could see nothing, but, while yet a hundred yards away, Bob was exulting, "A catch this time sure enough."

Dogs and horses all were inspired. The Terrible

Billy, the Dog That Made Good

Turk, realizing his importance, breasted his way to the front, and the rumbling in his chest was grand as an organ. Ahead, behind, and all around him, was Silly Billy yapping and tumbling.

There was the carcass, rather "high" now but untouched. The place of the trap was vacant, log and all were gone; and all around were signs of an upset, many large tracks, so many that scarcely any were clear, but farther on we got the sign most sought, the thirteen-inch track of a monster Grizzly, and the bunch on the right paw stamped it as Reelfoot's trail.

I had seen the joy blaze in Yancy's eye before, but never like now; he glowed with the hunter's heat, and let the dogs run free, and urged them on with whoops and yells of "Sic him, boys!" "Ho, boys!" "Sic him!" Not that much urging was needed, the dogs were possessed of the spirit of the day. This way and that they circled, each for himself. For the Bear had thrashed around a while before at length going off. It was Croaker that first had the real trail. Big Ben was there to let the whole world know, then Thunder indorsed the statement. Had it been Plunger that spoke the rest would have paid no heed, but all the pack knew Thunder's voice, and his judgment was not open to question. They left their devious different

Billy, the Dog That Made Good

tracks, and flocked behind the leader, baying deep and strong at every bound, while Turk came hurrying after and Silly Billy tried to make amends in noise for all he lacked in judgment.

Intoxicating moments those for all the hunt. However civilized a man may be, such sounds and thoughts will tear to tatters all his cultured ways and show him up again a hunting beast.

Away we went, the bawling pack our guides. Many a long detour we had to make to find a horseman's road, for the country was a wilderness of rocky gullies, impenetrable thickets, and down timber, where fire and storm had joined to pile the mountain slope with one dead forest on another. But we kept on, and before an hour the dinning of the pack in a labyrinth of fallen trees announced the Bear at bay.

No one who has not seen it can understand the feelings of that hour. The quick dismount, the tying of the nerve-tense horses, the dragging forth of guns, the swift creep forward, the vital questions, "How is he caught? By one toe that will give, and set him free the moment that he charges, or firmly by one leg?" "Is he free to charge as far as he can hurl the log? or is he stalled in trees and helpless?"

Creeping from trunk to trunk we went, and once

the thought flashed up, "Which of us will come back alive?" Oh, what a din those dogs were making! Every one of them was in that chorus. Yapping and baying, high and low, swaying this way and that, which meant the Bear was charging back and forth, had still some measure of freedom.

"Look out now! Don't get too close!" said Yancy. "Log and all, he can cover fifty feet while you make ten, and I tell you he won't bother about the dogs if he gets a chance at the men. He knows his game."

THE FIERY FURNACE AND THE GOLD

There were more thrills in the woods than the mere sounds or expectations accounted for. My hand trembled as I scrambled over the down timber. It was a moment of fierce excitement as I lifted the last limbs, and got my first peep. But it was a disappointment. There was the pack, bounding, seething, yelling, and back of some brush was some brown fur, that was all. But suddenly the brushwood swayed and forth rushed a shaggy mountain of flesh, a tremendous Grizzly—I never knew one could look so big—and charged at his tormentors: they scattered like flies when one strikes at a gathered swarm.

But the log on the trap caught on a stump and

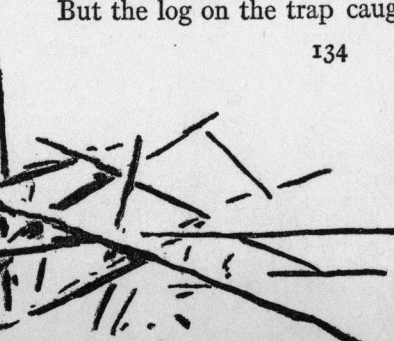

Billy, the Dog That Made Good

held him, the dogs surged around, and now my view was clear.

This is the moment of all in the hunt. This is the time when you gauge your hounds. This is the fiery furnace in which the metals all are tried. There was Old Thunder baying, tempting the Bear to charge, but ever with an eye to the safe retreat; there was Croaker doing her duty in a mere announcement; there were the greyhounds yapping and nipping at his rear; there in the background, wisely waiting, reserving his power for the exact proper time, was the Terrible Turk, and here and there, bounding, yapping, insanely busy, was Silly Billy, dashing into the very jaws of death again and again, but saved by his ever-restless activity, and proud of the bunch of Bear's wool in his teeth.

Round and round they went, as Reelfoot made his short, furious charges, and ever Turk kept back, baying hoarsely, gloriously, but biding his time for the very moment. And whatever side Old Thunder took, there Turk went, too, and Yancy rejoiced, for that meant that the fighting dog had also good judgment and was not over-rash.

The fighting and baying swung behind a little bush. I wanted to see it all and tried to get near, but Yancy shouted out, "Keep back!" He knew the habits of the Bear, and the danger of coming

Billy, the Dog That Made Good

into range. But shouting to me attracted the notice of the Bear, and straight for Bob he charged.

Many a time before had Yancy faced a Bear, and now he had his gun, but perched on a small and shaky rotten log he had no chance to shoot, and swinging for a clearer view, upraised his rifle with a jerk—an ill-starred jerk—for under it the rotten trunk cracked, crashed, went down, and Bob fell sprawling helpless in among the tumbled logs, and now the Grizzly had him in his power. "Thud," "crash" as the trap-log smote the trees that chanced between; and we were horror-held. We had no power to stop that certain death: we dared not fire, the dogs, the man himself, were right in line. The pack closed in. Their din was deafening; they sprang on the huge haired flanks, they nipped the soggy heels, they hauled and held, and did their best, but they were as flies on a badger or as rats on a landslide. They held him not a heart-beat, delayed him not a whit. The brushwood switched, the small logs cracked, as he rushed, and Bob would in a moment more be smashed with that fell paw, for now no human help was possible, when good old Thunder saw the only way—it meant sure death for him—but the only way. Ceased he all halfway dashing at the flank or heel and leaped at the great Bear's throat. But one

Hanging to the Big Bear's face, flapping like a rag, was Silly Billy

swift sweep of that great paw, and he went reeling back, bruised and shaken. Still he rallied, rushed as though he knew it all must turn on him, and would have closed once more, when Turk, the mighty warrior Turk, the hope and valor of the pack, long holding back, sprang forward now and fastened, gripped with all his strength—on the bear? *No*, shame of shames—how shall I say the truth? *On poor old Thunder*, wounded, battered, winded, downed, seeking to save his master. On him the bulldog fastened with a grip of hate. This was what he waited for, this was the time of times that he took to vent his pent-up jealous rage—sprang from behind, dragged Thunder down to hold him gasping in the brushwood. The Bear had freedom now to wreak revenge; his only doughty foeman gone, what could prevent him? But from the reeling, spieling, yapping pack there sprang a small white dog, not for the monster's heel, not for his flank, or even for his massive shoulder forging on, but for his face, the only place where dog could count in such a sudden stound, gripped with an iron grip above the monster's eye, and the huge head jerking back made that small dog go flapping like a rag; but the dog hung on. The Bear reared up to claw, and now we saw that desperate small white dog was Silly Billy, none else, hanging on with all his might and weight.

Bob scrambled to his feet, escaped!

The huge brute seized the small white body in paws like stumps of trees, as a cat might seize a mouse he seized, and wrenched him quivering, yes, tore his own flesh wrenching, and hurled him like a bundle far aside, and wheeling for a moment paused to seek the bigger foe, the man. The pack recoiled. Four rifles rang, a long, deep, grating snort, and Reelfoot's elephantine bulk sank limp on the storm-tossed logs. Then Turk, the dastard traitor Turk, with chesty gurgle as a war-cry, closed bravely on the dead brute's haunch and fearlessly tore out the hair, as the pack sat lolling back, the battle done.

Bob Yancy's face was set. He had seen it nearly all, and we supplied the rest. Billy was wagging his whole latter end, shaking and shivering with excitement, in spite of some red stained slashes on his ribs. Bob greeted him affectionately: "You Dandy. It's the finish that shows up the stuff a Bear-dog is made of, an' I tell you there ain't anything too good in Yancy's Ranch for you. Good old Thunder has saved my life before, but this is a new one. I never thought you'd show up this way."

Billy, the Dog That Made Good

"And you," he said to the Turk, "I've just two words for you: 'Come here!'" He took off his belt, put it through the collar of the Terrible Turk, led him to one side. I turned my head away. A rifle cracked, and when at length I looked Yancy was kicking leaves and rubbish over some carrion that one time was a big strong bulldog. Tried in the fire and found wanting, a bully, a coward, a thing not fit to live.

But heading all on the front of Yancy's saddle in the triumphal procession homeward was Billy, the hero of the day, his white coat stained with red. His body was stiff and sore, but his exuberant spirits were little abated. He probably did not fully understand the feelings he had aroused in others, but he did know that he was having a glorious time, and that at last the world was responding to the love he had so bounteously squandered on it.

Riding in a pannier on a packhorse was Old Thunder. It was weeks before he got over the combined mauling he got from the Bear and the bulldog, and he was soon afterward put in honorable retirement, for he was full of years.

Billy was all right again in a month, and when half a year later he had shed his puppy ways, his good dog sense came forth in strength. Brave as a Lion he had proved himself, full of life and energy,

Billy, the Dog That Made Good

affectionate, true as steel, and within two years he was leader of the Yancy pack. They do not call him "Silly" now, but "Billy, the pup that made good."

V
Atalapha, a Winged Brownie

V
Atalapha, a Winged Brownie*

*I have always loved the Brownies so much, and so earnestly wished to believe in them, that I have taught myself to do so, and I want others to have that same pleasure. It is worth your while looking up some good old books (not new ones) to learn, if you wish to do so, just what a Brownie is. I think you could find that all the good reliable authorities like Grimm, Andersen, etc., agree that the Brownie is a shy two-legged elfin wearing a fur cloak, standing about a thumb high in his silent stockings, though he never does stand that way.

He is distinguished from other two-legged dwarves by his sharp-pointed ears and his sense of humor. He gets his living by dancing over the treetops in the woods on moonlight nights and differs from other fairies in being quite friendly to man. He dwells in a cave or hollow tree, hiding all day, and either sleeps all winter underground or steals away to some warm country; though without feathers, he is blessed with marvellous powers of flight. Besides which, he can talk without making a noise, is invisible at will in the moonlight, and has many wonderful powers that we children understand perfectly, but are beyond the comprehension of our wisest grown-ups.

THE TWINS

THE Beavers had settled on the little brook that runs easterly from Mount Marcy, and built a series of dams that held a succession of ponds like a wet stairway down the valley, making a

Atalapha, a Winged Brownie

break in the forest that gave the sky a chance to see its own sweet face in the pools below.

They were peaceful folk, the Beavers, and many of the shy little hush-folk, that the fairy books tell of, were glad and welcome to harbor and revel in the pleasaunce these water-workers had created. Thus it came about that the cool green aisle of the timber land was haunted, in the Beaver vale east of Marcy.

The Rose Moon glowed on the pine-robed mountain. The baby Beavers were learning to slap with their tails, and already the chirring in high places told of young birds grown and lusty. The peace of the forest was abroad, for it was calm and cool in the waning light.

The sun sets thrice each day in Marcy Vale: First, when it drops so far that the tall timber on the western slope steeps all below in a soft green shade; this the sunset of the forest; again, when the great rugged breast of Mother Marcy blocks out all light from the trees; this is the sunset of the Mountain; and last, when the western world rim receives the light orb, the mountain's brow turns red for a moment, then ashy pale; this is the sunset of the world. In a little while then all is dark, the sun peoples go to sleep and the hush peoples of shadowland have now their day.

The portrait of a Brownie

Atalapha, a Winged Brownie

The sunset of the forest had given the signal to robin and tanager to begin their vesper song. The sunset of the mount had issued the dew-time call that conjures out of caves and hollow trees the smallest of the winged Brownie folk, whose kingdom is the twilight and whose dance hall is high above the treetops.

Now they come trooping down the open aisle above the Beaver ponds. Skimming and circling on lightning wing, pursuing each other with shouts that to them seemed loud and boisterous, though to us they were merely squeaks and twitters too thin and fine for any but the sharpest ears.

Up and down the waterway they dart, playing, singing, hunting. Yes, hunting, for this is the time and place of the evening meal and the prey they catch and eat is—as befits such dainty coursers of the air—the butterflies of the night. And when one of those great fluffy things went fluttering by, some two or three of the Brownie throng would cease pursuing gnats and gauze flies, to have a riotous breakneck speeder after the moth, and rending its fat body in the air among them, they scattered its feathers to the wind and its framework to the ground.

There was a fixed order for the coming of the winged ones, an etiquette, not written, but ob-

Portrait of a Brownie

served: just as the smaller folk come earliest in any procession, so the lesser Elfins in their scores were first to arrive.

In half an hour the black-faced Brownies came in hundreds, and the air over the tranquil Beaver ponds was like that of a barnyard whose swallow colony is strong.

The third sunset came and went. The shades of night were sweeping up from the east. The robins alone were singing in the gloaming, when beautiful Borealis in his red and yellow robes skimmed down the mountain-side and joined the jolly pirouetting host that sang and circled in the upper shades.

A little later long-winged Serotinus skimmed into the crowd, to be the advance courier of the last and royalest of them all, that clad in frosted sable furs swooped in on ample wings. Biggest, strongest, rarest of the folk of Shadowland, the king of his kind, the chief of the winged Brownies, and yet we sordid blind ones have no better name for him than Hoary Bat.

Darting up and down the waterway, chasing the fat moths and big game of the night, noctua, samia, lachnosterna, or stripping their bodies of legs and wings to devour the soft parts in air, the great Bat flew, first of the royal house to come. Sometimes skimming low over the waters, sometimes shooting skyward above the trees, sometimes spinning up and down, faster than any of its lesser kin. One not gifted with night eyes would have marvelled to learn that in all this airy wheeling and speeding she, for it was a Queen Bat, carried a heavy burden. Clinging to her breast were two young Bats, her offspring. They were growing fast and already a heavy weight; but none who marked only the mother's flight would have guessed that she was so trammelled and heavy laden.

Up and down the fairway of the water she skimmed, or high above the trees where roam the bigger flyers of the night, till she had caught and eaten her fill, then after another hovering drink at the Beaver pond she left the almost-deserted flyway and soaring over the treetops, she made up the mountain-side to her home den, a knot-hole in a hollow maple too small to be entered by Marten or Hawk or any creature big enough to do her harm.

THE SCHOOLING OF A BROWNIE

As June, the Moon of Roses, passed, the young Bats grew apace. They were full furred now, and their weight so great that the mother left them in the den in the hollow branch each time she went forth seeking food. Now she brought back the bodies of her prey, moths and June-bugs; for the young were learning to eat solid food, and when their mother came home after the evening hunt, they would meet her at the door with a soft chirring of welcome, spring on the food she brought, and tussel with each other for the pieces.

Two meals a day, or rather each night, is a rule of the Bat life—one in the evening twilight, and again in the morning twilight. And twice each day the mother stuffed them with food, so they grew and grew. The difference of their dispositions was well marked now. The lesser brother was petulant and a little quarrelsome. He always wanted the June-bug that had not been given him, and paid little heed to the warning "chirr" that his mother sometimes gave to stop him scrambling after his brother's portion. But the bigger brother was not easily provoked; he sought for peace. What wonder that the mother found it pleasanter to stroke and lick the big one's fur than to be chittered at by the little one.

June went by, July the Thunder Moon was half gone, when a great event took place. The young had been growing with wonderful rapidity. Though far from being as heavy as the mother yet, they were nearly as long and had a wing stretch that was fully three-quarters of hers. During the last few days they had dared to sit on their home branch outside of the den, to wait for mother with the eatables. Each time they saw her coming their well-grown wings fluttered vigorously with excitement, and more than once with such power that the young bodies were lifted almost off their feet; surely the time had come for the great experiment. Instead of giving them the food that evening, the mother Bat kept a little way off.

Holding the body of a cockchafer, she alighted on a branch, and when the hungry little ones pursued her clamoring, she kept just out of reach, and continued on to the end of the branch. The little ones scrambled after her, and just as they reached the prize she launched into the air on her wings. The Big Brother was next her. He had been reaching for the food; the suddenness of the move upset him. He lost his hold and in a moment was falling through the air. He gave a little screech, instinctively spread out his wings and flapped very hard. Then lo! instead of falling, he went fluttering forward, and before he knew it, *was flying!*

It was weak and wabbly, but it was flight. Mother was close at hand, and when he seemed to weaken and failed to hold control, she glided underneath and took his weight upon her back. Wheeling, she mounted with strong, sturdy strokes. Soon again he was back to the home den and his maiden flight was over. It was three days before Little Brother would take his flight. And many a scolding his mother gave him before he could be persuaded that he really had wings to bear him aloft, if only he would try to use them.

From this time on the twins' real life began. Twice nightly they went flying with Mother to the long wet valley through the timber, and though at first they wearied before they had covered thrice the length of the Beaver ponds, their strength grew quickly, and the late Thunder Moon saw them nearly full grown, strong on the wing, and rejoicing in the power of flight. Oh! what a joy it was, when the last streak of light was gone from the western world rim, to scramble to the hole and launch into the air—one, two, three—Mother, Brother, and Little Brother to go kiting, scooting, circling, sailing, diving, and soaring—with flutter, wheel, and downward plunge. Then sharp with hunger they would dart for the big abounding game—great fat luna moths, roaring June-bugs, luscious cecropias, and a thousand smaller game were whizzing and flitting on every side, a plenteous feast for those with wings of speed. One or two small moths they seized and gobbled in mid-air. Then a fat June-bug came booming by and away

Chasing the booming June-bug

Atalapha, a Winged Brownie

went the youngsters twittering with glee, neck and neck, and Mother hovering near. Within half a pond length they were up to him, and pounced and snapped, Little Brother and Big Brother. But an unexpected difficulty arose. The June-bug was so big and round, and clad in such hard-shell armor, that each time the young Bats pounced and snapped, their little jaws could get no hold but sent the bug rebounding, safely speeding.

Snap, snap, snap, went the little Bats, but it was like terriers snapping at an Armadillo, or kittens at a Turtle. For the June-bug kept his legs tight tucked and all the rest was round and hard. "Snap" went Brother at his head and "snap" went Little Brother at his tail. They nearly bumped into each other, but the booming bug escaped, and Little Brother chittered angrily at every one.

Then the Mother Bat came skimming by and said in Bat language: "Now, children, watch me and see how to manage those big hard things you cannot bite." She swooped after the roaring bug, but making no attempt to use her teeth she sailed over, then in a twinkling curled her tail with its broad flap into a bag, and scooped the June-bug in. Her legs helped to close the net; a quick reach back of the supple neck and the boomer

Atalapha, a Winged Brownie

was seized by the head. Her hind feet clutched it firmly, a few quick movements of her jaws: the wing cases, the armored legs and horns went down rattling into the leafage, and the June-bug's body was like a chicken trussed for eating, cleaned of all but the meat.

Calling to the twins with a twittering squeak, she took the fat lump in her teeth and flew onward and upward, still calling. Then as they labored in pursuit, she rose a little and dropped the big luscious prize.

Away went Brother, and after went Little Brother, in pursuit of the falling food. It fell straight, they darted in zigzags. Again and again they struck at it, but could not hold it. It was surely falling to the ground, where it would be lost, for no Frosted Bat would eat food from the ground. But Mother swooped, and with her tail scooped the round thing in again.

Once more she flew to the higher level above the trees. Again she called to the brothers to try their powers. And as the fat body dropped a second time they resumed their eager zigzags. A little screech of joy from Little Brother announced that he had scooped the body, but he lost his wing balance, and dropped the June-bug to recover himself. It had not fallen twenty feet before

Atalapha, a Winged Brownie

Brother dashed under sideways and up, then twittered in needle tones of joy, for he had won the prize and won it in fair play. The old Bat would have eaten it on the wing, but the little ones were not yet steady enough for that, so they flew to a tall tree, and to a top branch which afforded a good perch, and there they revelled in the spoils.

THE UNDOING OF LITTLE BROTHER

The Thunder Moon was worthy of its name. Night after night there were thunderstorms that prevented the Bats going out to hunt, and the hardship of hunger was theirs, for more than once they had to crouch in the home den while the skies and trees shivered in thunder that shook down drenching streams of rain. Then followed a few clear days and nights with growing heat. Little Brother, always petulant, chittered and crooned in querulous notes, but Brother and Mother bore it all silently. The home was surely very close, but it was a safe refuge. At last Little Brother would stand it no longer. The morning hunt was over, that is the second meal, the east was showing a dawning. All three had huddled in the old safe home, but it got closer and hotter, another blazing day was coming, and Little Brother, in spite of warning chitters from his mother and bead-eyed

Atalapha, a Winged Brownie

wonder of his brother, crawled out of the den, and hung himself Bat fashion, heels up, under a thick and shady spruce bough close at hand.

Mother called once or twice, but he answered her only with an impatient grunt, or not at all. He was very well pleased to find it so much cooler and pleasanter under this bough than in the den, though in truth the blinding sun was far from agreeable.

The brightness and the heat grew and the bird voices mostly died away. But there was one that could be heard in sun or shadow, heat or twilight, the loud *"Jay, jay"* of the Bluejay, the rampant, rollicking, mischief bird, the spy and telltale of the woods.

"Jay, jay!" he screamed, when he found a late fledgling in the nest of a Vireo and gobbled the callow mite as its parents wailed around. *"Jay, jay, too-rootel!"* he chortled as he saw a fat grasshopper left on a thorn by a butcher bird who believed in storing food when it was plenty. But the Jay polished off the dainty, and hopped gayly to a cleft tree into which some large insect had buzzed. The Jay tapped with his bill; an angry buzz gave warning.

"Nay, nay!" said the blue terror, and lightly flitted to a tall fir out of reach of the angry hornets.

Atalapha, a Winged Brownie

Here his keen eyes glancing around caught a glimpse of a brownish-looking lump like an autumn leaf or a moth cocoon.

"*Took, took,*" murmured the Jay. "What is that?" It hung from the lower side of a limb. The Jay hopped just above it. The slight jarring of his weight caused two tiny blinky eyes to open, but the sunlight was blinding, the owner was helpless, and with one fell blow of his sharp bill the Bluejay split its skull. The brown form of the Bat shook in the final throe, fell from the perch and was lost to view, while the Bluejay croaked and "*he he'd*" and went on in the rounds of his evil life.

That was the end of Little Brother.

His mother and brother knew he was killed, but they could see little of it in the brightness; they were sure only of this: they never saw him again.

But a man, a good naturalist, was prowling through the woods that day with trout rod in hand. It was too hot to fish. He was lying under a tree in the shade when the familiar voice of the Bluejay sounded above him. He saw nothing of the bird. He knew nothing of its doings overhead, but he did know that presently there fluttered down a beautiful form, the velvet and silver

Atalapha, a Winged Brownie

clad body of a Great Northern Bat, and when the wings had ceased to flutter, a closer glance showed that the skull was split by a blow from some sharp instrument. But the rare specimen was little harmed; he gladly took it to an honored resting-place. He had no answer to the riddle, but we know it for the working out of the law: Obedience is long life.

ATALAPHA'S TOILET

Atalapha, the Big Brother, now lived alone with his mother, learning many things that were needful to his life success; being taught by her, even as she was taught by her mother, chiefly through the power of example, developing so fast that he was full grown before the waning of the Thunder Moon, and was far advanced for his age in all the wise ways of Bats.

One of the first lessons was the making of his toilet, for the winged Brownies are exquisitely clean in their person. This was the way of his washing; After dipping once or twice in the water so the lower fur was dripping wet he would fly to some well-known roost and hanging by first one foot then by the other, would comb his fur all over with the thumb that grows on each wing bend; and then with finer applications of his teeth

and tongue, every part was dressed and licked as
carefully as a cat might dress her coat. And last
his wings are rubbed and massaged inside and out.
He would lick and pull at the membrane, and
stretch it over his head till every part was cleared
of every speck of dust, and the fur slick and clean
and fluffy soft.

He knew how to take the noisy June-bug in the
scoop net; how to snap the small but juicy May-
flies and mosquitoes in his mouth and cut their
wings with his side teeth. He could seize, strip,
pluck, bone, and eat a noctua or a snowy mænas
without changing his line of flight. He knew that
a polopias was a hardshell and a stinger to be let
alone; some young Bats have lost their lives
through not knowing what a deadly creature is
this steel-blue mud-wasp. He knew that the
woolly luna, the fluffy samia, with her Owl-eyed
wings, or the blazing yellow basilona cannot be
scooped, but must be struck from above and de-
winged. So also with the lightning hawkmoth,
the royal citheronia, and the giant cecropia, the
hardest of all to take, the choicest food in the air.
He learned to keep away from the surface of the
Beaver pond when the great trout were jumping,
and he had discovered the wonderful treat that
one may eat hovering in front of a high honey-

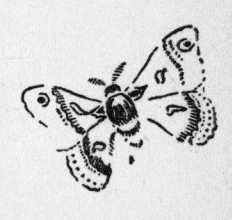

Luna

Atalapha, a Winged Brownie

suckle when honey, pollen, and smokeflies mixed made a thick delicious food that was a new sensation. He knew the booming hoot of the Horned Owl and the screech of the early Pigeon Hawk. He could dart at full speed, without touching, through an opening but little wider than one wing. He could comb his left side with his right thumbnail. He learned to enjoy teasing the great clumsy Nighthawks; and when he saw one spreading its enormous gape to close on some fat basilona, he loved to dart between and in a spirit of mischief and sport to bear the coveted morsel away. All Great Northern Bats are marvellous on the wing, but Atalapha was a marvel among the young of his kind. He rejoiced in the fullness of his speed. He gloried in the strength of his wings, and—shall I tell it?—he became a little puffed up. Because he pleased his mother, and was a little abler than his mates and had taken with credit the first steps in the life journey, he reckoned himself a very important being; and he thought he knew it all. He had an awakening.

THE COMING OF THE BRIDEGROOMS

With the closing of the Thunder Moon Atalapha found himself not only independent of his mother, but also that she, in yet a larger sense, was becoming independent of him. He was her equal in size, and though they kept the same den, they came and went more as they listed, often alone. Sometimes they did not meet at all in the hunt.

With the opening of the Red Moon another great change began. The mother left the den earlier, left it sometimes as soon as the first shadows had fallen on the forest, skimmed far away, he knew not where, came home later, would sometimes go out in the middle of the night, which is not the custom of the Bat, would leave in the morning too early for the morning meal, and come back perhaps near sunrise, tired but excited. She had not now the burden of nursing her young, and she filled out in flesh, her fur fluffed softer than velvet, and its rich brown, too, was frosted with silver tips that shone like skifts of snow. Her eyes grew bright and her cheeks, once so flat and thin, puffed out in rounded shape of health and vigorous desire. Some great change was setting in, and its first effect was to separate the mother from the son.

It was on the third or fourth day of the growing change, on toward the time of morning meal—the little star blink, when none but the faintest stars are blotted out—Atalapha and his mother had not yet aroused, when a strange sound came whinnying through the calm, clear air. It was new to Atalapha and gave him no special thrill; but on his young mother it acted like a spell. She scrambled to the doorway and launched into the night with a long, warbling, high-pitched, "Hoooooo!" Atalapha lost no time in following; and then in the starry night it seemed that every star had a quavering voice, was singing a soft, long, "Hee-ooo," "hee-ooo!" in strains so high that human ear may hear them not, in notes so soft and quavering that surely these were the Brownie bugles blowing. From everywhere and nowhere came the strains. But darting up and out Atalapha realized that the air was full of Bats, not the little black-faced things like those he had scornfully hunted with all summer, but Great Northern Bats like himself and his mother, and yet not just the same for they were bigger, stronger, more richly clad, like folk of his race, but nobles beside whom he was of common kind.

What were they? Whence came they? Why should they sing as he never had heard his people sing? How beautiful and big and strong they looked. What wondrous turns they made in air. And as he gazed he saw a pair come swooping by. One great handsome fellow with wings of eighteen span, and fur like flame, when it whistled red or yellow in the wind. And the smaller one! Could

The flittering Brownie host in the moonlight

Atalapha, a Winged Brownie

it be? Yes! Atalapha could not mistake that shoulder band of white, but he flew yet closer, and saw with a strange little feeling of loss that the silver collar-bearer *was his mother*. She sailed and swooped in close companionship with the big splendid fellow. Atalapha knew it not, but this was his father come back to his bride.

THE GREAT SOUTHERN TREK

The early meal was caught and eaten in the air; and when the beams of the Little Morning whitened the eastern screen, the new Bat host began to hide itself for the day. Many went to old familiar nooks; some went to the homes of their wives; but many who had brought their mates with them on this the great national wedding trip were left to roost in the hemlock tops, chance hollow limbs, or crannies in the rock.

Atalapha was already at home when the door darkened, and in came his mother with her big handsome mate. What a strong look he had, and how his coat did shine. Probably he did not realize that Atalapha was his child, for he showed his teeth when the young Bat came too near the little mother, and by a warning hiss sent him to the far side of the den. He did no more than this, but Atalapha was afraid. The young Bat did not clearly understand, but he realized that this was no longer his home, that the bond of the family was broken. So he came no more, but sought a den for himself, feeling as a child might feel when suddenly dropped from being mother's pet, for whom the world was made, to being a poor little outcast. At best now he belonged to the low circle of the young Bats whose powers were not yet formed. There was no joy in Atalapha's lone cell those days; but it was the beginning of life for him. He learned that he was a very unimportant person, and must begin at the bottom and go it all alone. This was his humiliation and his awakening.

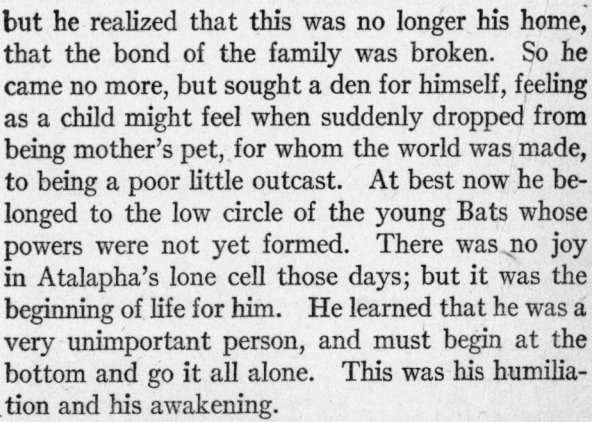

The love dance of the Bats held sway supreme through the first few days of August, then, though its opening rapture waned, the Red Moon was a Honey Moon to its end. And when the Hunting Moon came on with shorter days and fewer kinds of game, a new unrest possessed the kindred of the silken wings.

Atalapha's parents would go for long excursions, swinging round the Marcy Mount, or sometimes a group of many friendly pairs would soar above the zone of the midnight game and circle high as though trying their wings in some new flight.

Then came a day of climax. It was in the early dawning; all had made their meal; Atalapha was

Atalapha, a Winged Brownie

flying with some new acquaintances of the younger set, when the soft singing "*hee-oo, hee-oo*" was taken up by all and trooping together from divers parts of the broad valley the long-winged coursers came. They swirled like smoke around the cliffs of Marcy, they careered in a body, then they began to ascend in a great sweeping spiral. At first in one long wreathing cloud, but later in two separate bodies, and those with eyes to see would have known that the upper swarm was wholly of males, the lower of female Bats.

The clamor of their calling made a thin, fine murmur in the upper air, but it was chilly there, their voices died away and with one impulse they turned to the south, as the mountain-top turned red, and flew and flew and flew.

Every male of that gathering was there: instinctively Atalapha had joined them, and they flew with steady, uncurveted flight in ever-lengthening procession, on and on, all day, ignoring the sun, heeding not the pang of hunger, till in the evening they straggled into a wood far to the southward, and rested in the trees a while before beginning to hunt for food.

The females, including the Little Mother, were left behind in the shadows about Mount Marcy to follow in another band when the males were well ahead, for such is the way of the Great Northern Bat.

For long after the arrival of the van, the he Bats came straggling into that welcome woods. The sun went down, the moon arose, and still belated wanderers drifted in, yet some there were that never came at all, that failed and fell by the way. But early, among those that never flagged, was Atalapha; he had attached himself to the big Bat that really was his father, and flying as he flew, he got the help of larger wisdom, for the old Bat changed his course to fit the air currents. He avoided a head wind. He sought for aiding blasts. He shunned the higher ridges that make fighting swirls of air. He neither speeded nor slacked nor sailed, but kept up the steady, slow flap, flap, flap that eats up miles and leagues and makes great headway with the least of drain. So passed the day on the first long trek, and Atalapha's travels and broader education had begun.

It was a night or two before the Bats were rested enough to continue their journey. And now they took shorter stages, for the frost fear that had come upon them in the north was goading them no longer. Now also they went at times by night. At last they reached the sea and followed the main shoreline with the land to the sunset side and the

Atalapha, a Winged Brownie

endless blue to the east. Thus after the waning of the Hunting Moon, the Leaf-falling Moon found them in a land where there is no falling of the leaf, where the trees are never bare.

Here in the groves of palms, where purple moths and radiant fireflies make a fairy scene by night, and nature's table is prepared and spread in every glade, Atalapha and the brethren of his race scattered each to seek a hunting for himself. In little groups at most they kept together, but no great crowds, for kings go not in companies nor princes in mobs. So also it is more often seen that the greatest of the Bats is alone.

Their consorts, too, came drifting south to the land of winters warm and never-failing food. But they also lived their own lives, and if by chance they met their mates among the palms, they passed as merely kinsmen with whom they had no feud.

NORTHWARD, HOME AGAIN

Where there is no winter there can be no wonderful spring. Only the land of dreadful cold can thrill our souls with the glad yearly miracle of bees and violets, where just a month before was snow and fiercest frost. Yet even in the home of palms and endless warmth the Spring Moon came with hidden power; not a mighty change to hold the eye, but a

Atalapha, a Winged Brownie

secret influence that reached all life. The northern songbirds now showed different plumes. The Wild Geese and the Cranes found things just the same in this land of winter sun and food, but a change had come inside. An unseen prompter persistent, reiterant, unreasoning, sang in Atalapha's heart: "*Away, away; up and away!*" So it sang in every heart, and the Bat host moved, as the prairie grass is moved by the unseen wind, headed all one way, turned in the moon of Wild Geese, moved in the month of Greening Grass, swinging northward with a common impulse, even as they had come to the south in the autumn.

They made neither haste nor speed; but straggling as before, in a larger company than before, it might have been noted that in the Blue Mountains many left the host that followed the white wailing line of the sea, and took another course, for these were summer dwellers in the far northwest.

Atalapha and his kindred of the pine woods kept on. Their nightly journey and their nightly meals covered the country as fast as it was won again by victorious spring. One night a sudden change of weather sent all the Bats into hiding-places, where they huddled together and in many cases became insensible from the cold. For three days they lay hid and seeming dead, but all revived when the sun grew warm again, and, reassembling, took up their nightly northern trek.

Bats have a strong homing instinct, and group after group dropped from the main route as they reached the first river valleys or mountains that had been their guides coming south. When at last the far green woods loomed up ahead, Atalapha felt the glad thrill of "Home again." But it was not so to be. His guides, the flying speeders, never halted. He was free, of course, but another impulse was on him. He had no conscious expression for the fact, but he felt that this was a range for lady Bats; now he was a big strong male.

Whither?

This is the rule of the tribe, grown up and established for reasons unknown, except that it works out well, that certain parts of the range are the homes of the females, the nurseries of the young. The males are supposed to go to the farther mountains, to higher uplands, and remain till the time of the great annual reunion, the nuptials of the tribe.

This is the law of the Bats. There are no stripes or heavy tolls assigned or penalty expressed for those who break it, excepting that the Bat who lives where he should not live is left by himself, shunned and despised by his kin.

Thus the Marcy Vale had no lasting hold on

Atalapha, a Winged Brownie

Atalapha, and he flew with the dwindling troop his father led, till the roaring Saranac was blue beneath their wings.

WINGS AND FRIENDSHIPS

Atalapha had been growing all winter. His father no longer looked so very big and strong; indeed the son had dimly felt that the once big Bat was shrinking strangely. He himself was two heads wider in his wing expanse, and the dull yellow of his body fur had given place to rich ochre and amber brown with a wonderful frosting like silvery snow bespread on banks of gold. But these things he neither knew nor thought of; his conscious pride was this: the speed and strength and tireless force of his glorious silken wings.

On his wings he took his prey.

On his wings he eluded the Owls and late-flying Hawks or climbing beasts that were his only foes.

On his wings he raced with his fellows or skidded and glided, pirouetted and curveted in the air, playing pranks with Owl and Nighthawk as a greyhound plays round a Bear.

The appetite for food might quickly flag, but the joy of speeding like a falling star while the tang wind of the dawn went whistling past his ears, the glory of a lightning swift career, with

Atalapha, a Winged Brownie

none propelling but his own strong nervous force, with quick coördinate life in every film and fibre of his frame, with exquisite sensibility of power and risk and change of breath, with absolute control of every part and move. What greater joy could there be to Old Mother Nature's slow-made perfect flying thing than this the perfect joy of perfect flight?

Men have so long been envious of the flies and birds for their mastery of the air and pictured theirs a life of Paradise restored, they have forgotten that there is still another higher creature, a being nearer to ourselves whose babes are born alive, whose brain is on a higher plane than that of any bird, whose powers perceptive are of exquisite acuteness, whose make-up is attuned to sounds and senses we labor hard to prove; that nature made many a blundering trial with the scaled and feathered folk, but all her finished summing up of flight she centred in this her favorite, the highborn, cave-born Bat, that clad in exquisite furs, mounted on silent silken wings, equipped with wonderful senses, has so long led his blameless life so near our eyes, and yet so little on our ken.

With strength in his body and courage in his heart, Atalapha now led the twilight host that sallied forth from hollow tree or bosky hemlock top. Each evening their routine was the same. After sunset the horde of smaller Bats, then with twilight the smaller group of the Great Northern Bats. Leaving their den, they flew first to the river, where they drank as they sped along. Then for half an hour they hawked and fed on insects taken on the wing. Last came a time of social play, racing, chasing, games of tag and touch-me-not, with others of a dangerous kind. One of these, a favorite in time of heavy heat, was shooting the chutes where the Saranac leaps over a rocky ledge to be forgotten in foam. The reckless youngsters of the Bat fraternity would drop for a moment in the arrowy flood above the fall, and as they were shot into the abyss, would ply their dripping wings and sail through the spray-mist to repeat the chute, perhaps. There was no lack of danger in the sport, and more than one that summer took the leap to be seen no more.

Still another game of hazard had a little vogue. In the Saranac were great grown Trout; at the rare times when the Northern Bats chanced out before the sun was wholly gone, these Trout would leap at flies that the lesser Bats were chasing, and more than once a Bat that ventured low was leaped at by the monster Trout and barely escaped.

Then in a spirit of daredevil did Atalapha skim

Atalapha, a Winged Brownie

low and tempt the Trout to leap. None but the largest would rise to such a bait. But rise they did, and nothing saved the king of the air from the king of the pool but a marvellous upward bound of lightning speed. There was no lack of excitement in it, but when at last a little Bat was caught and gulped by a Trout of arrowy swiftness, and Atalapha himself had the skin ripped from his tail tip, the sport of trouting lost its charm.

Atalapha's den was now a knot-hole in an oak. The doorway was a tight fit as every cave-dweller desires it, but inside was ample room and every comfort that shape and sheltered place could give. But on a luckless day it occurred to an unscrupulous Flicker with defective property instincts that he could improve this hole by enlargement and convert it to uses of his own. So after listening to his nagging tap, tap, tap, all one day, and seeing the hole get unpleasantly large, Atalapha was forced to seek another den.

The place of his choice was not unlike the first, but the entry and den both were larger. Yet the former was too small to admit a Red Squirrel, and the Bat moved in.

Next morning when he returned from his early meal and was going off to sleep, he was aroused by a peculiar scratching. Then the hole was darkened and in came a great furry creature with big, black shiny eyes. At first it filled the Bat with fear, for there was no escape. But it was only a gentle mother Flying Squirrel looking for a nursery den. To a being of such exquisite sense power as a Bat, much knowledge comes without a sound or a visible sign, and in some such hidden way a something told Atalapha, "Be not afraid. This gentle, soft-furred, big-eyed creature will never do you harm."

Thus it was that Atalapha and Fawn-eyes came to share the den, and when the babies of the Flying Squirrel came, they found a sort of foster-brother in the Bat. Not that he fed or tended them, but each knew the other would do him no harm; both kinds loved to be warm, and they snuggled together in the common den, in closeness of friendship that grew as the season passed.

THE WINGED TIGER AND THE UNKNOWN DEATH

"*Hoo-hoo-ho-hooooo!*" A deep, booming sound—it came filling all the valley. Atalapha heard it with a scornful indifference. Fawn-eyes heard it with a little anxiety. For this was the hoot of the Great Horned Owl, the terror of the woods, the deadly, perhaps the deadliest, enemy of Bat and Flying Squirrel. Both had heard it before,

Atalapha, a Winged Brownie

and many times, but now it was so near that they must be prepared to face and in some way balk the flying death, or suffer hunger till it passed.

"*Hoo-hoo-hooooooooo!*" it came yet nearer. Old Fire-eyes and his mate were hunting in this valley now. It behooved all lesser revellers to heed their every move, and keep in mind that the grim and glaring tiger of the pines might any moment be upon them. Atalapha asked no odds but clear sky-way. Fawn-eyes had little fear except for her brood, and for herself asked only a thicket of lacing boughs. And both went forth as the shadows fell.

Then came a rare and wrenching chain of ill events! Fawn-eyes was doing her best to swoop across a twenty-five-foot space, when the huge and silent enemy perceived her, and wheeled with lightning dash to win the prey. But she reached the trunk and scrambling round took another flying leap for the next tree, hoping to gain safety in some mass of twigs, or, safer still, a hollow trunk that was not far away. But the Owl was quick, and wheeling, diving, darting, was ever coming closer. The swooping of his huge bulk, the vast commotion of his onset, caught Atalapha's attention. He came flying by, out of curiosity perhaps, and was roused to find first that his enemy was astir and next that

Atalapha, a Winged Brownie

the victim pursued was Fawn-eyes, his friend and den mate. It takes but little to make a Bat swoop for a foeman's face, even though he turn before he strike it, and the combination brought the big Bat like an arrow at the Owl's great head. He ducked and blinked, and Fawn-eyes reached the hollow tree, to scramble in and hide in a far small crack, the Owl in close pursuit. Again Atalapha swooped at the big bird's head, rebounding as Old Fire-eyes ducked, but rebounding—alas! right into the very claws of the second Owl who had hurried when she heard the snapping of her partner's bill. And down she struck! Had Atalapha been ten times as big he would have been riddled, crushed, and torn; but his body, small and sinewy, was over-reached. Not the claws but the heel had struck, and drove him down, so he dropped into the hollow tree, and scrambling quickly found a sheltering cranny in the wall.

Now was there a strange state of siege. Two huge Owls, one out, one in the hollow trunk, and in two lesser passages the Bat and the Flying Squirrel. The Owls had never lost sight of their prey, but they could not get their claws into the holes. Again and again they forced in one armed toe, but the furred ones crouching in their refuges could by shrinking back keep just beyond the

Atalapha, a Winged Brownie

reach of that deadly grip. Sometimes the Owl, failing to reach with claw, would turn his huge face to the place, snap his bill, and glare with those shiny eyes or make the tree trunk boom with his loud *"Hoo-hoo-hoo-ho!"*

Sometimes one of the monsters went off hunting. But always one stayed there on guard; and so the whole night passed. It was only when the sunrise was at hand that remembrance of their own unfed, unguarded nestlings took the Owls away; and so the siege was ended.

There were other friendships and other hazards in the life of Atalapha. Many of the male community were good fellows, to meet and pass in friendly evening flight. His father, now quite small it seemed, was of the brotherhood, meeting and passing or ceding little courtesies of the road as is the way of Bats; but he was a comrade, nothing more.

Some of these Bats lived in little groups of two or three, but most had a single cell where they slept. The little Black and Brown-faced Bats might roost in swarms, but the Great Northern Courser of the night more often dens alone. This was the habit of Atalapha, except during that brief summer time that he shared his home with Fawn-eyes.

The perils of his life were first the birds of prey,

Atalapha, a Winged Brownie

the silent, lumbering Owls; the late, or very early, flying Falcons; the Weasel, or the Red Squirrel that might find and enter his den while he slept; the Trout that might leap and catch him as he took his drink a-wing. And still worse than these the deadly Acarus that lodges in the Bat's deep fur. It is sure that the more the Bats harbor together in numbers, the more they are plagued by the Acarus. Yet there is a remedy. Instinct and example were doubtless the power that had taught him, for Atalapha clearly knew that when some Acarus lodged in his fur and made itself felt as a stinging tickly nuisance, the only course for him was a thorough hunt. Hanging himself up by one foot, he worked with the other, aided by his jaws, his lips, his tongue, and the supple thumbs on either wing.

There was no part of his body that he could not reach; in him the instinct of cleanliness was strong, so he never suffered vermin in his fur. When, as it chanced, through no fault of his the den became infested, there was but one remedy, that was *move out*.

Yet one other peculiar menace was there in the lives of the Saranac community. Far up on the higher waters one of the big human things that cannot fly had built a huge nest. Across the river, too,

Atalapha, a Winged Brownie

it had made a thing like a smooth Beaver dam and, Beaver-like, it had cut the trees for a wide space around. All this was comprehensible, but there was another strange affair. The two-legged thing had built a huge round nest of stones in the side of a hill and then when it was lined with tree trunks, it glowed by night with *the red mystery*, and strange fumes came pouring out, ascending to the sky in a space which changed with the wind. There was a weird attraction about this high place of different air. Bats would flutter near the furnace as it glowed by night and sent an upward wind of heat and pungent smell. Insects came, it is true, attracted by the light, but surely the Bats did not come for that, as there were plenty of insects elsewhere. Perhaps it was to taste the tingling, dangerous vapors that they came, just as some men find pleasure in teasing a coiled rattler or lingering barely out of reach of a chained and furious Bear. Some Bats came pursuing their lawful prey, and if they chanced to be flying low might enter the outer edge of the deadly gas before they knew, for it had no form or hue.

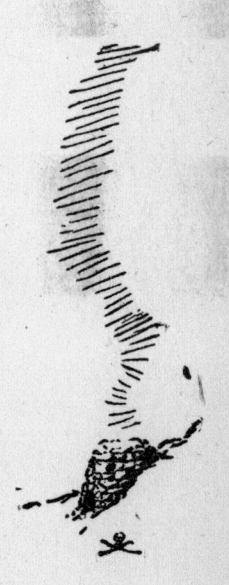

Atalapha plunged right into the vapor of the lime kiln once. He went gasping, sputtering through, nearly falling, but was able to sustain his flight till his breath came back, and slowly he recovered. Others were less lucky, and more than one of the birds of prey had learned to linger near the fiery kiln, for feathered things as well as Bats were often so stupefied by its fumes that they became an easy prey.

If it had any naming in the memory of the Bats, it was the Place of the Unknown Death.

ATALAPHA WOUNDED AND CAPTIVE

A good naturalist who found Bats worthy of his whole life study has left us a long account of a Bat roost where ten thousand of the lesser tribes had colonized the garret of a country dweller's home. It was in a land of flies, mosquitoes, and many singing pests with stings, but all about the house was an Eden where such insects were unknown. Each Bat needs many hundred little insects every night, what wonder that they had swept the region clear.

Slow-moving science has gathered up facts, and deciphered a part of the dim manuscript of truth that has in it the laws of life.

We know now that typhoid, malaria, yellow fever, and many sorts of dreadful maladies are borne about by the mosquitoes and the fly. Without such virus carriers these deadly pests would die out. And of all the creatures in the woods there is none that does more noble work for man

Atalapha, a Winged Brownie

than the skimming, fur-clad Bat. Perhaps he kills a thousand insects in a night. All of these are possibly plague-bearers. Some of them are surely infected and carry in their tiny baleful bodies the power to desolate a human home. Yes! every time a Bat scoops up a flying bug it deals a telling blow at mankind's foes. There is no creature winged or walking in the woods that should be better prized, protected, blessed, than this, the harmless, beautiful, beneficent Bat.

And yet, young Haskins of the Mill, when his uncle gave him a shotgun for his birthday, must need begin with practice on these fur-clad swallows of the night that skimmed about the milldam when the sun went down behind the nearer hills.

Again and again he fired without effect. The flittering swarm was baffling in its speed or its tortuous course. But ammunition was plentiful, and he blazed away. One or two of the smaller Bats dropped into the woods, while others escaped only to die of their wounds. The light was nearly gone from the western sky when Atalapha, too, came swooping down the valley about the limpid pond. His long, sharp wings were set as he sailed to drink from the river surface. His unusual size caught the gunner's eye, he aimed and fired. With a scream of pain the great Bat fell in the stream, and the heartless human laughed triumphant, then ran to the margin to look for his victim.

One wing was useless, but Atalapha was swimming bravely with the other. He had nearly reached the land when the boy reached out with a stick and raked him ashore, then stooped to secure the victim; but Atalapha gave such a succession of harsh shrieks of pain and anger that the boy recoiled. He came again, however, with a tin can; the wounded Bat was roughly pushed in with a stick and carried to the house to be shut up in a cage.

That boy was not deliberately cruel or wicked. He was simply ignorant and thoughtless. He had no idea that the Bat was a sensitive, highstrung creature, a mortal of absolutely blameless life, a hidden worker, a man-defender from the evil powers that plot and walk in darkness, the real Brownie of the woods, the uncrowned king of the kindly little folk of Shadowland; and so in striking down Atalapha the fool had harmed his own, but the linking of his life with the inner chain of life was hidden from him. Cruelty was far from his thoughts; it began with the hunting instinct, then came the desire to possess, and the gratification of a kindly curiosity—all good enough. But the methods were hard on the creature caught. The boy pressed his nose against the close wire netting and stared at the wet and trembling prisoner. Then the boy's little sister came, and gazed with big blue eyes of fear and wonder.

"Oh, give it something to eat," was her kind suggestion. So bread, for which the wounded one had no appetite, was pushed between the bars. Next morning of course the bread was there untouched.

"Try it with some meat," suggested one; so meat, and later, fish, fruit, vegetables, and, lastly, insects were offered to the sad-faced captive, without getting any response.

Then the mother said: "Have you given it any water?" No, they had never thought of that. A saucerful was brought, and Atalapha in a fever of thirst drank long and deeply, then refreshed he hung himself from a corner of the cage and fell asleep. Next morning the insects and all the fresh meat were gone; and now the boy and his sister had no difficulty in feeding their captive.

THE WINGS THAT SEE

Atalapha's hurt was merely a flesh wound in the muscle of his breast. He recovered quickly, and in a week was well again. His unhinging had been largely from the shock, for the exquisite nervous sensibilities of the Bat are perhaps unequalled in the animal world, how fine none know that have not been confronted with much evidence. There was once, long ago, a cruel man, a student of natural history, who was told that a Bat has such marvellous gift of nerves, and such a tactile sense that it could see with its wings if its eyes were gone. He did not hesitate to put it to the proof, and has left a record that sounds to us like a tale of magic.

There was sickness in the small settlement, and the doctor calling, learned of the children's captive. He knew of Spallanzani's account and was minded to test the truth; but he was not minded to rob a fellow-being of its precious eyesight. He could find other means.

Opening the cage, he seized the fur-clad prisoner, then dropping deftly a little soft wax on each eyelid, he covered all with adhesive plaster so that the eyes were closed, absolutely sealed; there was no possibility of one single ray of entering light. And then he let the captive fly in the room. Strong once more on the wing, Atalapha rose at once, in wavering flight, then steadied himself and, hovering in the air, he dashed for the ceiling. But a moment before striking he wheeled and skimmed along the cornice, not touching the wall,

Atalapha, a Winged Brownie

and not in seeming doubt. The doctor reached out to catch him, but the Bat dodged instantly and successfully. The doctor pursued with an insect net in hand, but the blinded Bat had some other sense that warned him. Darting across the room, he passed through the antlers of a Deer's head, and though he had to shorten wing on each side, he touched them not. When the pursuing net drove him from the ceiling, he flew low among the chairs, passing under legs and between rungs at full speed, with not a touch. Then in a moment of full career near the floor he halted and hovered like a humming-bird before the tiny crack under the door, as though it promised escape. All along this he fluttered, then at the corner he followed it upward, and, hovering at the keyhole, he made a long pause. This seemed to be a way of escape, for the fresh air came in. But he decided that it was too small, for he did not go near, and he certainly did not see it. Then he darted toward the stove, but recoiled before too close. The roaring draft of the damper held him a moment, but he quickly flew, avoiding the stovepipe wire, and hovered at another hairlike crack along the window.

Now the doctor stretched many threads in angles of the room and set small rings of wire in the narrow ways. Driven upward from the floor, the

Atalapha, a Winged Brownie

blinded prisoner skimmed at speed along the high corners of the room, he dodged the threads, he shortened wing and passed in full flight through the rings, and he wheeled from every obstacle as though he had perfect vision, exact knowledge of its place and form.

Then, last, the doctor gave a crucial test. On the table in the middle of the room he set a dish of water and released a blue-bottle fly. Every one present was cautioned to keep absolutely still. Atalapha was hanging by his hind feet from a corner of the room, vainly trying to scratch the covering from his eyes. Presently he took wing again. The dead silence reassured him. He began once more his search for escape. He made a great square-cornered flight all around the door. He traversed at a wing length the two sides of the sash, and then inspected the place where the cross-bars met. He passed a mouse hole, with a momentary pause, but hovered long at a tiny knot-hole in the outer wall. Then reviving his confidence in the silence of the room, he skimmed several times round and, diving toward the pan, *drank as he flew.* Now the fly that had settled on the wall went off with a loud hum. Instantly Atalapha wheeled in pursuit. It darted past the Deer's antlers and through the loops and zigzag threads round here and there, but not for long. Within half the room's length the fly was snatched in full career. Its legs and wings went floating away and the body made a pleasant bite of food for the gifted one.

What further proof could any ask, what stronger test could be invented? The one with the wonderful wings was the one with the tactile power that poor blind man gropes hard for words to picture, even in the narrow measure that he can comprehend it.

Tired with the unwonted flight, Atalapha was hanging from the wall. His silky seal-brown sides were heaving just a little with the strain. The butterfly net was deftly dropped upon him; then with warm water and skilful care the plasters and wax were removed, and the prisoner restored to his cage, to be a marvel and to furnish talk for many a day as "the Bat that could see with his wings."

Then in the second week of captive life there was a change: the boy came no more with coarse lumps of food, the sister alone was feeder and jailer, and she was listless. She barely renewed the water, and threw in the food, taking little note of the restless prisoner or the neglected cage. Then one day she did not come at all, and next day after hasty feeding left the door unlocked. That night Atalapha, ever searching for escape, trying every wire and airhole, pushed back the door, then skimmed into the room, and through an open window launched out into the glorious night again upon his glorious wings, free! free! free! And he swooped and sailed in the sweet fresh air of the starry night, and sailed and soared and sang.

And who shall tell the history of his bright young jailers at the mill? Little is known but this: the pestilence born of the flies alighted on that home, and when the grim one left it there were two new mounds, short mounds, in the sleeping ground that is overlooked by the wooden tower. Who can tell us what snowflake set the avalanche arolling, or what was the one, the very spark which, quenched, had saved the royal city from the flames. This only we know: that the Bats were destroying the bearers of the plague about that house; many Bats had fallen by the gun, and the plague struck in that house where the blow was hardest to be borne. We do not know. It is a chain with many links; we have not light to see; and the only guide that is always safe to follow in the gloom is the golden thread of kindness, the gospel of Assisi's Saint.

Atalapha, a Winged Brownie

ATALAPHA MEETS WITH SILVER-BROWN

The Thunder Moon was passing now. Atalapha was well and strong as ever, yes, more than ever before. He was now in his flush of prime. His ample wings were longest in the tribe, his fur was full and rich; and strong in him was a heart of courage, a latent furnace of desire. Strange impulses and vague came on him at times. So he went careering over the mountains, or fetching long, sweeping flights over the forest lakes from Far Champlain to Placid's rippling blue.

The exuberant joy of flight was perhaps the largest impulse, but the seeking for change, the hankering for adventure were there.

He sailed a long way toward Marcy Mount one night, and was returning in the dawning when he was conscious of nearing a place of peril. A dull glow in the valley ahead—the Unknown Death. And he veered to the west to avoid that invisible column of poison, when far to the east of him he heard a loud screeching, and peering toward the broad band of day that lay behind the eastern hilltops, he saw a form go by at speed with a larger one behind it.

Curiosity, no doubt, was the first motive to draw him near, and then he saw a Bat, one of his own

Atalapha, a Winged Brownie

kind, a stranger to him and of smaller finer make than his robust comrades on the Saranac. Its form brought back memories of his mother, and it was with something more than passing sympathy he saw she was being done to death by a bird of prey. It was early, but already the ravenous Chicken-hawk was about and haunting a place that had yielded him good hunting before. But why should a Bat fear the Chicken-hawk? There is no flyer in the sky that can follow the Great Hoary Bat, but follow he did, and the Bat, making wretched haste to escape and seeming to forget the tricks and arrowy speed of her kind, was losing in an easy race. Why? Something had sapped her strength. Maybe she did not know what, maybe she never knew, but her brain was reeling, her lungs were choking, she had unwittingly crossed the zone of the Unknown Death; and the Hawk screeched aloud for the triumph already in sight.

The fierce eyes were glaring, the cruel beak was gaping, the deadly talons reached. But the stimulus of death so near made the numbed Bat dodge and wheel, and again; but each time by a narrower space escaped. She tried to reach a thicket, but the Hawk was overcunning and kept between. One more plunge, the victim uttered a low cry of despair, when *whizz* past the very eyes of the

Atalapha, a Winged Brownie

great Hawk went a Bat, and the Hawk recoiled before he knew that this was another. Flash, flap, flutter, just before his eyes, and just beyond his reach, came the newcomer full of strength and power, quicker than lightning, absolutely scorning the slow, clumsy Hawk, while Silver-brown dropped limply out of sight to be lost in a hemlock top.

Now the Hawk was roused to fury. He struck and dived and swooped again, while the Bat skimmed round his head, flirted in his face, derided him with tiny squeaks, and flouted the fell destroyer, teasing and luring him for a while, then left him far away as the Sea-gull leaves a ship when it interests him no longer.

There was no deep emotion in the part the big Bat played, there was no conscious sex instinct, nothing but the feeling of siding with his own kind against a foe, but he remembered the soft velvet fur of Silver-brown as he flew, and still remembered it a little when he hung himself up for his day sleep in the hollow he felt was home.

THE LOVE FIRE

The Red Moon rose on Saranac, and with it many a growing impulse rose to culmination. Atalapha was in his glorious prime; the red blood

Atalapha, a Winged Brownie

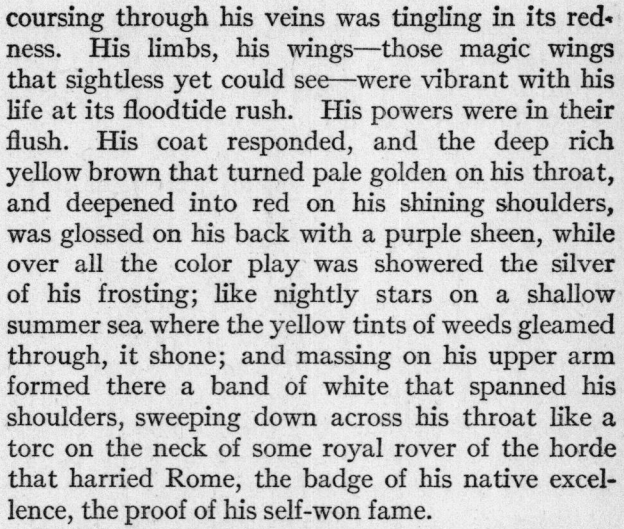

coursing through his veins was tingling in its redness. His limbs, his wings—those magic wings that sightless yet could see—were vibrant with his life at its floodtide rush. His powers were in their flush. His coat responded, and the deep rich yellow brown that turned pale golden on his throat, and deepened into red on his shining shoulders, was glossed on his back with a purple sheen, while over all the color play was showered the silver of his frosting; like nightly stars on a shallow summer sea where the yellow tints of weeds gleamed through, it shone; and massing on his upper arm formed there a band of white that spanned his shoulders, sweeping down across his throat like a torc on the neck of some royal rover of the horde that harried Rome, the badge of his native excellence, the proof of his self-won fame.

Rich indeed was his vestment now, but his conscious pride was the great long-fingered pulsatory wings, reaching out to grasp huge handfuls of the blue-green night, reaching, bounding, throbbing, as they answered to the bidding of the lusty heart within. Whether as a bending bow to hurl himself, its arrow, up toward the silent stars, or to sense like fine antennæ every form or barricade, or change of heat or cold, or puff of air, yes, even hill or river far below, that crossed or neared his unseen

Atalapha, a Winged Brownie

path. And the golden throat gave forth in silver notes a song of joy. Sang out Atalapha, as every sentient being sings when life and power and the joy of life have filled his cup brimful.

And he whirled and wheeled, and shrilled his wildest strain, as though his joy were rounded out complete.

How well he knew it lacked!

Deep in his heart was a craving, a longing that he scarcely understood. His life, so full, so strong, was only half a life; and he raced in wanton speed, or plunged like a meteor to skim past sudden death for the very pride and glory of his power. And skirling he spieled the song that he may have used as a war song, but it had no hate in its vibrant notes; it was the outbursting now of a growing, starkening, urging, all-dominating wish *for some one else*. And he wheeled in ever-larger lightning curves; careering he met his summer mates, all racing like himself, all filled with the fires of youth and health, burning and lusty life, that had reached a culmination—all tingling as with some pungent, in-breathed essence, racing, strenuous, eager, hungry, hankering, craving for something that was not yet in their lives, seeking companionship, and yet when they met each other they wheeled apart, each by the other shunned, and circling, yet voyaging in the upper air they went, drifting, sailing alone, though in a flock, away to the far southwest.

Fervent in the fervent throng and lightning swift among the flashing speeders was Atalapha in his new ecstatic mood. He had perhaps no clear thought of his need and void, but a picture came again and again in his mind, the form of a companion, not a lusty brother of the bachelor crew, but the soft, slight form of Silver-brown. And as his feelings burned, the impulse grew and his fleet wings bore him like a glancing star, away and away to the valley where ten nights back he had seen her drop as the Death Hawk stooped to seize her.

Star! red star of the Red Moon nights!

Star blazing in the sky, as a ruddy firefly glowing in the grass, as a lamp in a beacon burning!

Oh! be the wanderer's star to-night and guide him to the balm-wine tree!

Oh! shine where the cooling draft awaits the fevered lips and burning!

The strong wings lashed on the ambient wind, and that beautiful body went bounding, swinging, bounding. High, holding his swift line, he swept o'er Saranac and on. Low, glancing like an arrow newly sped, he traversed Pitchoff's many-shouldered peak. Like a falling star he dropped to Placid's broad blue breast and made across the waving forest heads.

For where? Did he know? For the upper valley of the river, for the place of the Unknown Death, for the woods, for the very tree in whose bosky top he had had the last, the fleeting glimpse of the soft little Silver-brown.

There is no hunger for which there is no food. There is no food that will not come for the hunger that seeks and seeks, and will not cease from seeking. Speeding in airy wheels in the early night, careering around the hemlock top as though it held, and had held these many days, the magnet that he had never realized till now—and many of his brethren passing near wove mystic traceries in the air; he sensed them all about, but heeded none —a compass for a compass has no message—when a subtle influence turned him far away, another power, not eyes nor tactile wings; and he wheeled with eager rush as one who sees afar a signal long awaited.

There! Yes! A newcomer of his race, of different form perhaps, and size and coat, but these were things he had no mind to see. This had a different presence, an overmastering lure, a speechless bidding not to be resisted, a sparkling of the

Atalapha, a Winged Brownie

distant spring to the sandworn traveller parched, athirst.

Now sped he like a pirate of the air. Now fled she like a flying yacht gold-laden, away, away, and the warm wind whistled, left behind. But the pirate surely wins when the prize is not averse to being taken. Not many a span of the winding stream, not many a wing-beat of that flight ere Atalapha was skimming side by side with a glorified Silver-brown. How rich and warm was that coat. How gentle, alluring the form and the exquisite presence that told without sounds of a spirit that also had hungered.

"*He-ooo, he-ooo, he-ooo!*" loud sang Atalapha in ecstasy of the love dream that came true.

"*He-ooo, he-ooo, he-ooo!*" and she sailed by his side. And as they sped the touch of lips or ears or wing-tips was their lover greeting, or tilting each away, as side by side they flew, their warm soft breasts would meet and the beating hearts together beat in time. The seeing wings supplied their comprehension in a hundred thrills, magnetic, electric, overwhelming. So they sailed in the blue on their bridal flight; so the hunger-mad joined in a feast of delight; so the fever-burnt drank at the crystal spring, for the moon that was full was the Red Love Moon, and it blazed on the brawling river.

Atalapha, a Winged Brownie

THE RACE WITH THE SWALLOWS

The fiercer the fire the faster it fades; and when seven suns had sunk on Marcy Vale, Atalapha and his bride, and the merry mated host that came that night from Saranac, were roaming in the higher winds with calmer flights and moods. The coursers of the night went often now alone. The ardor of the honeymoon was over, and strange to tell with the dulling of that fire the colors of their coats dulled, too.

August the Red Moon passed, and according to their custom the Bats prepared to go, like ancient pilgrims, in two great flights, the males in one, their consorts in a different later company.

Atalapha had seen no more of Silver-brown during the last week than he had of many others, and the law was easily obeyed. She was living with her kind, and he with his.

Then came again the stirring times when the nights turned cold. At last there was a nip of frost, and a great unrest ran through the Bat community. Next morning, after feed time, Atalapha made not for his lurking place, but wheeled toward the open, and after him the flittering host sailing and circling high. They were not dashing in feverish excitement as a month before, but wheeling upward as with a common purpose, so when the great spiral flock had soared so high that it was like smoke reflected in the river far below, its leader wheeled in a final wheel on the air current that suited him best, all followed, and their journey was begun. A troop of Swallows came fleet-winged from the north, and so the two swarms went together.

It seems impossible for two swift creatures not actually companions or mates to travel the same road long without a race.

At first each Bat that happened to be near a Swallow took care not to be left behind. But the interest grew, and not half the first little valley was crossed before the rivalry between chance Swallow and chance Bat had grown till the whole Swallow army was racing the whole army of Bats, and Atalapha was matched with a splendid fellow in steely blue, whose wings went whistling in the wind.

Away they sped, keeping the same air level and straggling out as the different individuals showed their different powers. Who that knows the merry, glancing Swallow can doubt that it must win? Who that has watched the Northern Bat could ever have a question? Yet the race was nearly even. There were Bats that could not hold their own with certain Swallows, and there were Swallows that strained very hard indeed to keep near the Bats. Both sped away at their swiftest pace. A second valley was crossed and then a low range of hills. Both armies now were strung out at full length, and yet seemed nearly matched. But there was one trick that the Swallows could not keep from doing, that was curveting in the air. The habit of zigzag flight was part of their nature. The Bats often do it, too, but now, with speed as their aim, they laid aside all playful pranks of flight, and, level-necked like a lot of Wild Geese, flapping steadily at a regular beat, beat, beat, dropping or rising as their sensitive feelings showed was wise when the air current changed, their wings went beat, beat, beat. Another valley crossed, Atalapha made better choice of the air levels, and his rival dropped behind. His kinsmen followed. The Swallows began to lose a little, then, losing ground, lost heart; and before another river had been passed the first of the Swallows had dropped behind the last of the Bats, and silken wings had beaten whistling plumes.

LOST ON THE WATER

Most migrants seek the sea if it be anywhere near their course, no doubt because of the great guide line of its margin. Down the Connecticut Valley they had sped, and were not far from the sounding shore when the leader of the Bats led his following into hanging quarters for the day.

They were a tired lot, especially the youngsters, whose first long flight it was, and when the evening meal hour came most of them preferred to go on sleeping. The night was waning, the morning was coming, when the leader roused the host, and all went out to hunt. The great game season was over and food was so scarce that the sun arose while many yet were hunting, and now it was time to be moving on the long south march. Turning the gold of his breast to the southward, Atalapha with his friends in long array behind went swinging easily down the valley to the sea, when a change of wind was felt, a chilly blast from the north arose. The leader soared at once to seek a pleasanter level, but found it worse, then sank so far that at last they were tormented with eddies answering to the contour of the hills, and flitting low, were surprised with a flurry of snow that sent them skurrying into sheltered places, where they hung and shivered, and so they passed the rest of that day and the night, after a slowly gathered meal.

The dawn time came, and the Bats were all astir, for the spirit of unrest was on them. The snow was gone and the weather mild, so they held their course

Atalapha, a Winged Brownie

till the crawling sea was far below them, and its foaming sandy shore was the line that guided their army now.

The day had opened fair, but they had not sailed an hour before the sky was darkened, a noisy wind was blowing in changing ways, and an overstream of air came down that was stinging, numbing cold.

Wise Bats know that the upper air may be warm when the world is cold, and Atalapha soaring led in a long, strong, upward slope, and on a warmer plane he sped away. But in a little while the world below was hidden in a flying spume of fog that was driven with whiteness, and in that veil the Bats again were lost: only the few strong flyers near him could be seen; but Atalapha sped on. He saw no landmarks, but he had a winged thing's compass sense. So he flew high above the veiled world, never halting or fearing—but on.

He would surely have kept the line and outflown the storm but for a strange mischance that brought him face to face with an ancient foe.

The mizzling fog and driving sleet had ceased for a little so that he could see some distance around. A few of his daily comrades were there, but among them flying also was the huge brown form of a Hawk. He was sailing and flapping by turns, and easily wheeling southward rather than moving by direct flight. But as soon as he saw the Bat so near he turned his cruel head with those hungry yellow eyes and made for him, with the certainty that here was an easy meal.

Atalapha was a little cold but otherwise fresh, and he eluded the onset with scarcely an effort, but the Hawk, too, was fresh. He swooped upward again and again, so the flight became a succession of zigzags. Then the fog and snow closed in. The Hawk made another pounce which Atalapha easily dodged with a swift upwheel that took him far from danger of those claws, but also, as it happened, into a thicker, chillier cloud than ever, and so far as he could see, he was alone in space. His other sense, the vision of his wings, was dulled by the cold; it told him that the enemy was not so far away, but that was all; and he sped in the white darkness of the mist, as fast as he could, away from the boding menace.

Still he went at his steady pace. He saw no more of the Hawk, but the fog and the snow grew heavier; then the wind arose and he followed, for he could not face it, and flew on and on. The day should have come in brightness, but the clouds were heavy above, so he sailed and sailed. Then when sure he was safe and would descend to rest, he lowered through the snow-laden wind to find that there was *nothing below but the sea*, heaving, expanding, appalling, so he rose and flew again for a long, long time, then he descended to find—the awful sea. He arose once more, flew on and on and on, and still on, but the sea was below him. Then the snowstorm ceased, the sky cleared off as the sun began to go down, and the Bat's little eyes could glance round and round to see nothing but heaving sea, no sight of tree or land or any other Bats, nothing but the dark, hungry waters. He flew, not knowing whither or why, the only guide being the wind now falling; he was no longer numbed with cold, but he was wearied to the very bone.

Yet the only choice was go on or go down, so he flapped and sailed as he had since the dawn, and when the favoring breeze died away he soared a little, hoping to find another helpful wind, and sailed with his worn, weary wings—sailed as the hunger pang weakened him—sailed, not the least knowing whither. Had he had the mind of another being, that thought might have struck him down, but his animal frame was strong, his vision of danger was small, and he sailed ever onward and on.

THE REMORSELESS SEA

An hour, and another hour, slowly passed; the sun had gone, the soft light that he loved was coming down, but his spirit was failing. He did not know where he was going, or whether he should turn and follow the sun till he dropped. As soon as the doubt came on him, he felt his strength go. He kept on, but it was a feeble flutter, with little direction. Surely now the sea would swallow him up, as it doubtless had done many of his fellows. His courage never really failed till now. His flight was drifting downward, when far behind he heard a strange loud cry, a sound of many voices, and a backward glance showed skimming low over the water a far-flung string of long-winged birds, smaller than Hawks, black and white, whistling as they flew. The instinct to save himself caused him to rise higher, but his flight was slow now, and the broad-fronted horde of ocean roamers came up and past him with a whirring and a whistling, to fade in the gloom to the south.

They had paid no heed to him, yet when they were gone they helped him. He did not know that these were Golden Plovers migrating. He did not know that they were headed for the ocean islands where winter never comes, but the force of their example was not lost. Example is the great teacher of all wild things, and spurred by the clamorous band, Atalapha took fresh heart and, following their very course, flapped on, wearily, hungrily, slowly

Atalapha, a Winged Brownie

for him, but on. The night wind followed the sun for a time, but Atalapha put forth a little of his feeble strength to rise till he found an upper breeze that was warm and would help him.

All day from earliest dawn he had flown, in the early part at least in peril of his life, not a bite had he eaten, but on and on he kept, not the swift, swooping flight of the arrowy Bat as he comes when the shadows fall on Saranac, but slowly flapping and low, like a Heron flying with heavy, flagging flight, without curvet, but headed with steady purpose, swerving not, and on.

Six hundred miles had he flown; his little breast was heaving, the rich dark fur was matted with the spray, the salt on his lips was burning, but on and on he flew.

Flap, flap, flap. There was no sound but the moan of the sea, nor sight for his eyes to rest on, nor hint that his magic wings could sense a place of refuge; but on and feebly on.

Flap—flap—flap—there was naught but the pitiless ocean, and the brave little heart was sinking, and yet on—on.

Flap—flap. His eyes were long dimmed. His wings were forgetting their captain, but on—on—in the wake of the Plovers, still on.

The All-mother, inexorable, remorseless always,

Atalapha, a Winged Brownie

sends, at least sometimes, a numb sleep to dull the last pang, and the wing-wearied flyer was forgetting—but on in a slow, sad rhythm that was surely near the end, when away out ahead in the darkness came a volume of sound, a whistling, the same as had passed him.

Like a thrill it ran through his frame, like food and drink it entered his body, and he bounded away at a better pace. He put forth his feeble strength and flew and flew. Then the clamor grew loud. A great shore appeared, and all along the strand were the Plovers running and whistling. Oh! haven! oh, heaven at last! Oh! rest. And he sailed beyond the sand, there flung outspread, shivered a little, and lay still.

The remorseless All-mother, the kindly All-mother, that loves ever best her strong children, came and stood over him. She closed his eyes in a deathlike sleep, she flirted the sand sedge over him, that no shore-mew nor evil creature of the sea might do him harm. So he slept; and the warm wind sang.

The All-Mother

THE BROWNIES OF THE BLOOD ROYAL

The sandflies fluttered over him and the Plovers whistled along the shore as he lay, when the sun

Atalapha, a Winged Brownie

arose, but the All-mother was kind, had blown the grass about him; it hid him from the hungry Gull and from the sun's noon rays. The little tide of mid-ocean rose on the beach but did not reach him in his deathlike sleep. The second tide had risen and gone, and the sun had sunk in the dark western waters before he stirred. He shivered all over, then slowly revived; the captain awoke, took anew the command of the ship—Atalapha was himself once more. He was conscious but weak, and burnt with a fervent thirst.

His wings were strong but bone-tired and stiff. Spreading them out, he rose with an effort. The water was there. He sailed over it and dipped his lips only to sputter it out. Why had he forgotten? Had not he learnt that lesson?

With parched and burning tongue he sailed inland. A broad, rocky pool was dragging down a fragment of the bright sky to contrast it with the dull ground. He knew this was right. He sailed and dipped. Oh, joy! Sweet, sweet water! Oh, blessed balm and comfort! Sweet and cool with recent rain! He drank till the salt was washed from his burning lips. He drank till the fever fled, till his body's pores were filled, till his wings were cool and moist, and now his brain was clear, and with strength renewed, he swept through the

Atalapha, a Winged Brownie

air, and about that pool found a plenteous feast—found food in a glad abundance.

* * * * * *

Who would follow his unheroic winter life in those isles of eternal summer? Or who will doubt the spring unrest that surely comes, though there be no vernalization of the hills? Or the craving for home and at last the bold dash on a favoring wind over ocean's broad, pitiless expanse, with the clamoring birds, and of his landing, not broken, but worn, in the pines of a sandy coast, and the northwest flight on the southeast wind, with his kin once more, till again ere the change of the moon he was back on the reaches of Saranac, chasing the fat noctuas, scooping the green darapsas, or tearing the orange tiger-moths that one time looked so big and strong to him?

You may see him if you will, along the pond above Haskins' mill; you will know him by his size and marvellous flight. You may see him, too, if you spend a winter in the Bermudas, for he loves to take that vast heroic flight just as an Eagle glories in the highest blue for the joy of being alone on the noblest plane of exploit.

Yet another thing you should know: If you seek the cool green forest aisles made by the Beaver

Atalapha, a Winged Brownie

pond east of Marcy you will marvel when the Winged Brownies come. They are there in merry hordes; the least come first, and quite late in the evening, if you watch, you will see a long-winged Bat in velvet fur of silver-brown with a silver bar on either shoulder. Still later in the season, if you have wonderful eyes, you may see flying with her two others of the royal blood, with orange fur and silver on the shoulders, only in their case the silver is complete and goes right across, exactly as it does on Atalapha.

VI
The Wild Geese of Wyndygoul

VI
The Wild Geese of Wyndygoul
THE BUGLING ON THE LAKE

WHO that knows the Wild Northland of Canada can picture that blue and green wilderness without hearing in his heart the trumpet "honk" of the Wild Geese? Who that has ever known it there can fail to get again, each time he hears, the thrill it gave when first for him it sounded on the blue lake in the frame of green? Older than ourselves is the thrill of the gander-clang. For without a doubt that trumpet note in springtime was the inspiring notice to our far-back forebears in the days that were that the winter famine was at end—the Wild Geese come, the snow will melt, and the game again be back on the browning hills. The ice-hell of the winter time is gone; the warm bright heaven of the green and perfect land is here. This is the tidings it tells, and when I hear the honker-clang from the flying wedge in the sky, that is the message it brings me with a sudden mist in the eyes and a choking in the throat, so I turn away, if another be there, unless that other chance to be one like myself, a primitive, a "hark back" who, too, remembers and who understands.

So when I built my home in the woods and glorified a marshy swamp into a deep blue brimming lake, with Muskrats in the water and intertwining boughs above, my memory, older than my brain, harked hungry for a sound that should have been. I knew not what; I tried to find by subtle searching, but it was chance in a place far off that gave the clue. I want to hear the honkers call, I long for the clang of the flying wedge, the trumpet note of the long-gone days.

So I brought a pair of the Blacknecks from another lake, pinioned to curb the wild roving that the seasons bring, and they nested on a little island, not hidden, but open to the world about. There in that exquisite bed of soft gray down were laid the six great ivory eggs. On them the patient mother sat four weeks unceasingly, except each

The Wild Geese of Wyndygoul

afternoon she left them half an hour. And round and round that island, night and day, the gander floated, cruised, and tacked about, like a war ship on patrol. Never once did the gander cover the eggs, never once did the mother mount on guard. I tried to land and learn about the nest one day. The brooding goose it was that gave the danger call. A short quack, a long, sharp hiss, and before my boat could touch the shore the gander splashed between and faced me. Only over his dead body might my foot defile their isle—so he was left in peace.

The young ones came at length. The six shells broke and the six sweet golden downlings "peeped" inspiringly. Next day they quit the nest in orderly array. The mother first, the downlings closely bunched behind, and last the warrior sire. And this order they always kept, then and all other times that I have knowledge of. It gave me food for thought. The mother always leads, the father, born a fighter, follows—yes, obeys. And what a valiant guard he was; the Snapping Turtle, the Henhawk, the Blacksnake, the Coon, and the vagrant dog might take their toll of duckling brood or chicken yard, but there is no thing alive the gander will not face for his little ones, and there are few things near his bulk can face him.

So the flock grew big and strong. Before three

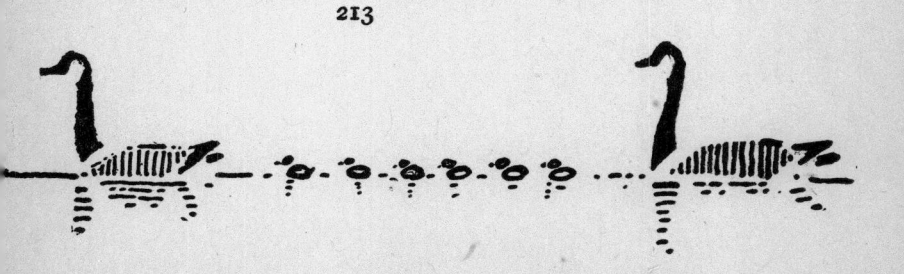

months they were big almost as the old ones, and fairly fledged; at four their wings were grown; their voices still were small and thin, they had not got the trumpet note, but seemed the mother's counterparts in all things else. Then they began to feel their wings, and take short flights across the lake. As their wings grew strong their voices deepened, till the trumpet note was theirs, and the thing I had dreamed of came about: a wild goose band that flew and bugled in the air, and yet came back to their home water that was also mine. Stronger they grew, and long and high their flights. Then came the moon of falling leaves, and with its waning flocks of small birds flew, and in the higher sky the old loud clang was heard. Down from the north they came, the arrow-heads of geese. All kinsmen these, and that ahead without a doubt the mother of the rest.

THE FIFTH COMMANDMENT

The Wild Geese on my lake turned up their eyes and answered back, and lined up on the lake. Their mother led the way and they whispered all along the line. Their mother gave the word, swimming fast and faster, then quacked, then called, and then their voices rose to give the "honk"; the broad wings spread a little, while they

The Wild Geese of Wyndygoul

spattered on the glassy lake, then rose to the measured "Honk, honk"; soaring away in a flock, they drifted into line, to join those other honkers in the Southern sky.

"Honk, honk, honk!" they shouted as they sped. "Come on! Come on!" they inspired each other with the marching song; it set their wings aquiver. The wild blood rushed still faster in their wilding breasts. It was like a glorious trumpet. But— what! Mother is not in the line. Still splashed she on the surface of the lake, and father, too— and now her strident trumpet overbore their clamorous "On, on! Come on!" with a strong "Come back! Come back!" And father, too, was bugling there. "Come back! Come back!"

So the downlings wheeled, and circling high above the woods came sailing, skirting, kiting, splashing down at the matriarchal call.

"What's up? What's up?" they called lowly all together, swimming nervously. "Why don't we go?" "What is it, mother?"

And mother could not tell. Only this she knew, that when she gave the bugle note for all to fly, she spattered with the rest, and flapped, but it seemed she could not get the needed send-off. Somehow she failed to get well under way; the youngsters rose, but the old ones, their strong leaders, had

The Wild Geese of Wyndygoul

strangely failed. Such things will come to all. Not quite run enough no doubt. So mother led them to the northmost arm of the lake, an open stretch of water now, and long. They here lined up again, mother giving a low, short double "honk" ahead, the rest aside and yet in line, for the long array was angling.

Then mother passed the word "Now, now," and nodding just a little swam on, headed for the south, the young ones passed the word "Now, now," and nodding swam, and father at the rear gave his deep, strong, "Now, now," and swam. So swam they all, then spread their wings, and spattered with their feet, as they put on speed, and as they went they rose, and rising bugled louder till the marching song was ringing in full chorus. Up, up and away, above the treetops. *But again*, for some strange reason, mother was not there, and father, too, was left behind on the pond, and once again the bugle of retreat was heard, "Come back! Come back!"

And the brood, obedient, wheeled on swishing wings to sail and slide and settle on the pond, while mother and father both expressed in low, short notes their deep perplexity.

Again and again this scene took place. The autumn message in the air, the flying wedges of their kin, or the impulse in themselves lined up

The Wild Geese of Wyndygoul

that flock on the water. All the law of ceremony was complied with, and all went well but the climax.

When the Mad Moon came the mania was at its height; not once but twenty times a day I saw them line up and rise, but ever come back to the mother's call, the bond of love and duty stronger than the annual custom of the race. It was a conflict of their laws indeed, but the strongest was, *obey*, made absolute by love.

After a while the impulse died and the flock settled down to winter on the pond. Many a long, far flight they took, but allegiance to the older folk was strong and brought them back. So the winter passed.

Again, when the springtime came, the Blacknecks flying north stirred up the young, but in a less degree.

That summer came another brood of young. The older ones were warned away whenever near. Snapper, Coon, and ranging cur were driven off, and September saw the young ones on the lake with their brothers of the older brood.

Then came October, with the southward rushing of the feathered kinds. Again and again that line upon the lake and the bugle sound to "fly," and the same old scene, though now there were a dozen

The Wild Geese of Wyndygoul

flyers who rose and circled back when mother sounded the "retreat."

FATHER OR MOTHER

So through the moon it went. The leaves were fallen now, when a strange and unexpected thing occurred. Making unusual effort to meet this most unusual case, good Mother Nature had prolonged the feathers of the pinioned wing and held back those of the other side. It was slowly done, and the compensating balance not quite made till near October's end. Then on a day, the hundredth time at least that week, the bugle sang, and all the marchers rose. *Yes! mother, too*, and bugling louder till the chorus was complete, they soared above the trees, and mother marshalled all her brood in one great arrow flock, so they sailed and clamoring sailed away, to be lost in the southward blue—and all in vain on the limpid lake behind the gander trumpeted in agony of soul, "Come back! Come back!" His wings had failed him, and in the test, the young's allegiance bound them to their mother and the seeking of the southern home.

All that winter on the ice the gander sat alone. On days a snow-time Hawk or some belated Crow would pass above, and the ever-watchful eye of

The Wild Geese of Wyndygoul

Blackneck was turned a little to take him in and then go on unheeding. Once or twice there were sounds that stirred the lonely watcher to a bugle call, but short and soon suppressed. It was sad to see him then, and sadder still as we pondered, for this we knew: his family never would come back. Tamed, made trustful by life where men were kind, they had gone to the land of gunners, crafty, pitiless and numberless: they would learn too late the perils of the march. Next, he never would take another mate, for the Wild Goose mates for life, and mates but once: the one surviving has no choice—he finishes his journey alone.

Poor old Blackneck, his very faithfulness it was that made for endless loneliness.

The bright days came with melting snow. The floods cut through the ice, and again there were buglers in the sky, and the gander swam on the open part of the lake and answered back:

"Honk, Honk, come back,
Come back. Come back!"

but the flying squads passed on with a passing "honk!"

Brighter still the days, and the gander paddled with a little exultation in the opening pond. How

The Wild Geese of Wyndygoul

we pitied him, self-deluded, faithful, doomed to a long, lone life.

Then balmy April swished the woods with green; the lake was brimming clear. Old Blackneck never ceased to cruise and watch, and answer back such sounds as touched him. Oh, sad it seemed that one so staunch should find his burden in his very staunchness.

But on a day, when the peeper and the woodwale sang, there came the great event! Old Blackneck, ever waiting, was astir, and more than wont. Who can tell us whence the tidings came? With head at gaze he cruised the open pond, and the short, strong honk seemed sad, till some new excitation raised the feathers on his neck. He honked and honked with a brassy ring. Then long before we heard a sound, he was bugling the marching song, and as he bugled answering sounds came—from the sky—and grew—then swooping, sailing from the blue, a glorious array of thirteen Wild Geese, to sail and skate and settle on the pond; and their loud honks gave place to softer chatter as they crowded round and bowed in grave and loving salutation.

There was no doubt of it. The young were now mature and they seemed strange, of course, but this was sure the missing mate: the mother had come

The Wild Geese of Wyndygoul

back, and the faithful pair took up their life—and live it yet.

The autumn sends the ordered flock afar, the father stays perforce on guard, but the bond that binds them all and takes them off and brings them back is stronger than the fear of death. So I have learned to love and venerate the honker Wild Goose whom Mother Nature dowered with love unquenchable, constructed for her own good ends a monument of faithfulness unchanging, a creature heir of all the promises, so master of the hostile world around that he lives and spreads, defying plagues and beasts, and I wonder if this secret is not partly that the wise and patient mother leads. The long, slow test of time has given a minor place to the valiant, fearless, fighting male; his place the last of all, his mode of open fight the latest thing they try. And by a law inscrutable, inexorable, the young obey the matriarch. Wisdom their guide, not force. Their days are long on earth and the homeland of their race grows wide while others pass away.

VII
Jinny. The Taming of a Bad Monkey

VII

Jinny. The Taming of a Bad Monkey

A DANGEROUS BRUTE

THE cage that arrived at Wardman's Menagerie was heavily bound with iron, and labelled *"Dangerous";* and when John Bonamy, the head-keeper, came up close to peep in, a hoarse "*Koff, koff,*" and a shock against the bars warned him that the label was amply justified. Through the grating his practised eye made out the dark visage of a Hanuman, or Langur Monkey, the largest and strongest of the kinds that come from India, a female, but standing over three feet high, and of bulk enough to be a dangerous antagonist, even to a man.

The other keepers gathered around, and the Monkey worked herself up into a storm of rage, leaping against the bars whenever one of the men

Jinny. The Taming of a Bad Monkey

came near enough to seem reachable. A scraper put in to clean up a little was at once seized in her paws, and mangled with her teeth. Keefe of the monkey house felt called on to take charge of things, and was peering in when suddenly a long, thin, hairy arm shot out and snatched off the goggles he was wearing, scratching his face at the same time, and putting him in an awful temper, which the merriment of the other men did nothing to allay.

The head-keeper had gone elsewhere, after giving instructions, but the noise and fuss brought him back. His trained ear detected signs of a familiar happening.

"You've got to remember they're human," he said, as he sent all the other keepers away and "sat down beside that crazy Monkey, to talk to her."

"Jinny," said he, giving her the first she-name that came handy, "now, Jinny, you and I have to be friends, and we will be as soon as we get better acquainted." So he kept on talking soothingly, not moving hand or foot, but softly cooing to her.

She was very ugly at first, but, responding to the potent mystery called personality, she gradually calmed down. She ceased snorting, and sat crouching in the filth at the back of the box, glowering with restrained ferocity, nervously clasping one skinny paw with the other. Bonamy did not mean

Jinny. The Taming of a Bad Monkey

to move for some time, but the wind lifted his hat, and as his hand flew up to seize it, the Monkey flinched, blinked, and again broke out in her sounds of animal hate.

"Oh-ho!" said he. "Some one has been beating you." Now he noticed the scars and certain slight wounds on her body; he remembered that she had crossed in a sailing vessel, and a measure of all that that meant came to him. He could imagine the misery of that long, long voyage, the fearful, ceaseless rolling, the terrible seasickness that so many monkeys suffer from, the shameful cruelty that he more than suspected, the bad food, and last the cramped and filthy cage before him. It was easy to guess the fact: the Monkey had had a horrible experience with men.

Bonamy was a born animal-man; he loved his work among them. He could handle and ultimately tame the most dangerous; and the more difficult they seemed, the more he enjoyed the task of winning them over. He could have controlled that Monkey in a day, but he had other things to attend to; so merely instructed the monkey-keeper to cover the filthy travelling coop with canvas and carry it to the hospital. Inside the big cage there it was partly opened; and at nearly every rap of the hammer the Langur gave a savage snort. Then

Jinny. The Taming of a Bad Monkey

from a safe place outside, a keeper pulled open the coop door.

Some animals would have dashed out at once, but Jinny did not. She crouched back, glaring defiantly from under her bushy moving brows, and seemed less inclined to come out now than when the coop was tightly nailed up.

Bonamy left her alone. He knew that it didn't do to hurry her. You can't be polite in a hurry, Lord Chesterfield says, and you must be polite to win your animals. Moreover, the story that the keeper read in her wounds showed that the human species had a black past to live down in Jinny's estimation.

She did not leave the coop all day. But that evening after sundown Bonamy peeped in, and saw her in the big cage washing her face and hands at the trough. Probably it was her first chance to be clean since she had left India. No doubt she had drunk what she needed, and now she glanced nervously about the place. The food supply she sniffed at, but did not touch; she walked gingerly around the ironwork, rubbed her finger on some fresh tar just outside the bars, smelt her finger, came back and drank more water, caught a flea on her thigh, then resumed her inspection of the bars. But she did not touch the food. Like

Jinny. The Taming of a Bad Monkey

ourselves, monkeys do not want to eat when they are all upset, they want a drink of water and quiet.

Next day she was perched up high, so the keeper put in his long hook to draw out the travelling coop. She sprang at him and raged against the bars. He tried to drive her back by prodding with the hook, but that only made her worse.

Bonamy had often warned his men against getting into a fight with the animals. "It does no good and only spoils our show." So Keefe came to him, grumbling: he "couldn't do nothing with that crazy Monkey." As soon as the two men entered the building Jinny sprang toward them, mad with rage; then Bonamy knew that Keefe had done more than he had owned up to. He sent him away and, standing very still, began to talk to the Monkey. "Now, Jinny," said he, "aren't you ashamed of yourself? Here, we want to be your good friends and help you, and this is how you go on!" It took fully ten minutes of that gentle talking and that strong, kind personality before the Monkey would listen to reason and get calm. She climbed up to the high shelf and sat there scowling, lifting her eyebrows and watching this big man, so different from the others she had met.

Realizing that the keeper had in some way incurred the Monkey's hate, he set about removing

Jinny. The Taming of a Bad Monkey

the dirty coop, and managed it after one or two little scenes, each one less violent than the last, but each guided by his rule never to scare any animal, never to hurt them, and always talk to them, *very softly*. He did not pretend that they knew what he said, but he felt they got the idea that he was friendly, and that was enough.

He soon found that it would not do to let Keefe tend her at all—the sight of that man was enough to set her crazy—so just because the taming promised to be a difficult job, Bonamy undertook it himself.

JINNY FINDS A NEW LIFE

After a week in quarantine Jinny was wonderfully improved, her fur was clean, her scratches healing, and she seemed less in terror of every approaching sound. Bonamy now decided that she was fit for the big show cage. There was a small trap cage on the highest point of her quarters, and watching till she was in that he pulled the string, then transferred the little cage and its inmate to the big outdoor place with over a dozen of other monkeys.

Of course she raged at the men during the removal. But they got her safely placed, and knew she would be quite a drawing card, for the public does love a noisy, fighting animal.

Jinny. The Taming of a Bad Monkey

As soon as she began to feel a little at home she charged at the other monkeys, sending them helter skelter and chattering to their highest perches, while she walked up and down, puffing out little snorts, raising and dropping her bushy eyebrows, and glaring defiantly at all the men outside.

The regular keeper came to feed her, and as usual went inside in spite of her angry threats. As soon as his back was turned she sprang and got him by the leg. He was badly bitten and she was hurt before she was driven off. But they knew now that it was no bluff, she was a "bad Monkey."

There seems to be a fascination about a thorough-paced villain, and Jinny was so bad that she was interesting. So yielding to an impulse, the big man with the strong hands and the soft heart settled down to his self-appointed task of bringing her "in line."

When he went to feed her she leaped up on a high perch, snorting, glaring, making faces, jumping up and down on all fours, daring him to enter. He was not looking for trouble, so he did not go in, but he was observing her keenly. One thing was sure: Jinny was no coward, and that was a great point; a brave animal is far easier to tame than a coward, as every Zoo-man knows.

He fed and watered the monkeys in that cage as

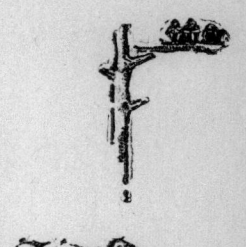

Jinny. The Taming of a Bad Monkey

well as he could from the outside, to avoid stirring up Jinny, but she kept drifting around to the edge of the cage where he happened to be, uttering a low, menacing sound, scratching her ribs with her little finger, jumping up and down, and occasionally dashing at the bars. She bullied all the other monkeys in the cage, too, but the man noticed that she had not really harmed any of them, even when she had good opportunity.

One morning before the public was in he was witness of an unusual affair: there was one very little Monkey that was terribly afraid of Jinny, and he usually kept one eye on her. But now he was at the front corner of the bars, wholly absorbed in an attempt to steal a banana from the next cage. He was so busy that for a moment or two he did not look around. Meanwhile Jinny had sneaked up softly, and now stood over him with her hands raised about six inches above his back. The little chap worked away unconsciously, barely reaching the banana with one finger, which he would bore into the fruit, then bring back to suck with gusto. At length, turning to look behind, he found he was trapped by his enemy.

In a moment he was a picture of abject terror. He crouched screaming in the corner of the cage, and Jinny, to the joy and surprise of the head-

Jinny. The Taming of a Bad Monkey

keeper, stood quite still, raised her hands a little higher, looked amused, he thought, and—let the victim go."

"Well," said he, "that settles it. I know she is not a coward and she is not cruel. She's *not* a bad Monkey at all. She's been abused, but she is all right and I am going to handle her before a month."

Then he began his old proven method, never scare her, move gently, go as often as he could, and always talk to her softly. At first when he came she would rush threateningly at the bars, then, finding that procedure barren of all interesting results, she gave it up in less than a week. But she would sit high on some perch and glare at him, scratching her ribs, puffing, and working her eyebrows. He used to joke her about it, as he phrased it, and in a fortnight could see he was winning the fight.

All this time there had been no thorough cleaning of the cage, only a "long-scraper" clean-out, so one morning he said: "I'll go in and scrub up." The boss warned him not to go. "That's a dangerous Monk," said he. "If she gets you by the neck, you are done."

But in he went. Jinny jumped up to her high perch and began snorting, jumping, and scratch-

Jinny. The Taming of a Bad Monkey

ing her ribs as usual. He kept one eye on her and talked to her all the time he was in, and nothing happened, but the boss warned him again. "You look out or she'll get you yet! I'll not be responsible if you go in there again!'

It was only a question of time and patience now, and Bonamy knew the business. Many visits, unvaried gentleness, soft talkings, little gifts of favorite food at each visit, and gradually resentment gave way to toleration, toleration to interest, and interest to attraction.

"I'll never forget the first time she let me scratch her head with a stick," said he. "I felt as proud as if I was a star batsman winning the pennant on a home run."

Thus she learned to look for his visits, and before the month was up Jinny and he became pretty good friends. His judgment of her was right: she had a fine character, was unusually intelligent, and only needed the chance he gave her. In her worst rampaging she had never hurt any of the little monkeys. She never seemed savage at women or children. She resented only the men. But now she was becoming quite tame even with them, except that she always hated Keefe, and the sight of a sailorman roused her to fury.

But her friendship for Bonamy grew daily; she

Jinny. The Taming of a Bad Monkey

would come running to meet him, and if he passed the cage without noticing her, she would jump up and down on all fours, scratching her ribs with her little finger, and giving a peevish, "Errr, errr." She was in good health now, and mentally as keen as a brier. She had more sense, the keeper used to say, than "some humans he could name." With her renewal of life and strength, and the total elimination of perpetual terror and sense of cruelty, she developed a most lively disposition. She was full of tricks that were partly due to her active brain and partly her physical energy. And strange to say, she also showed that at bottom hers was a most affectionate nature. As Bonamy said, she turned out to be the best Monkey he ever handled. She was worth more than a Lion to draw the public. She could take the crowd away from the Elephant and keep them, too, and seemed to have a pride in it, she was so nearly human. There was not an animal in the Zoo that the keepers thought as much of as Jinny. They learned to count on her now to "swing the whole thing" when there was a special day for school children.

THE SOUL OF A MONKEY

Three months had barely gone since Jinny came, and though not an important animal judged by

Jinny. The Taming of a Bad Monkey

the catalogues of dealers, there is little doubt that she was the head-keeper's favorite. It was not wholly because of his own triumph in converting her from an outlaw into the "most lovable Monkey he ever knew," but because back of her bright dark eyes there really seemed to be a personality almost human; keenly alert, deeply affectionate, and Bonamy's morning walk to the office took him invariably now to call first on Jinny.

One morning he was late in arriving. There was a crowd of visitors around the cage as he went by. Every few minutes a small outburst of applause or laughter showed that some of the animals there were making hits with the audience, and he was not surprised to catch a glimpse of Jinny busy at her usual antics. He had indeed guessed that it was her crowd, for she had more drolleries than all the rest put together. She used to walk a tight rope after chalking her feet with a piece of chalk given her at first in play, but she was taught to use it, and later learned to chalk the end of her nose at the same time, to the joy of the multitude. Her other specialty was to stand on her head near the front bars, catch hold high up with her hind feet, then swing herself up bodily sidewise till her front feet had hold far above her hind ones; then repeat the movement till she had

Jinny. The Taming of a Bad Monkey

rolled herself all the way to the top, reversing the loops to climb down again.

In spite of printed warnings, some woman passed under the barrier and reached forward to pull the tail of another Monkey who was crouching with his back to the public, and came so near that Jinny snatched her hat off, and putting it on her own head, continued to perform, and drew still louder rounds of applause from the crowd. There can be no doubt that she appreciated the applause, for it was noticed that she always did best for a crowd. Most monkeys have a human side, but Jinny was unusually gifted that way, and the head-keeper had a personal interest in her, so that now he went to his office with a sense of personal pride.

Jinny meanwhile played her lively pranks to a lively audience. Small boys threw peanuts which she ignored, for her cheeks were already bulging with them, and grown-ups threw bonbons which she promptly rescued from the other monkeys, for she was the largest in the cage and had the well-earned reputation of being a dangerous fighter. Every one but the owner of the lost bonnet was convulsed with joy as she dissected it bit by bit, and spat out the pieces that she tore from the trimming. Then responding to the tenth encore, she

Jinny. The Taming of a Bad Monkey

began her back somersaults up the iron-work. Just as she was drawing herself up with her breast tight against the bars, a coarse but foppish-looking man, yielding to some incomprehensible, diabolic impulse, reached out a long sword cane and stabbed the monkey in the groin. With a scream of pain she fell to the ground, and at once the scene was changed. A wave of fear and dismay sent all the lesser monkeys chattering to the high perches. The near onlookers were shocked and were loud in their cries of "Shame!" while those behind were struggling to find out what had happened.

Why do men do these cruel things? That horrible beast had actually stabbed that little Monkey for the mere pleasure of inflicting pain.

After the first scream Jinny had fallen, then she dragged herself to the far end of the cage, where she sat moaning, with her hands on the wound. The crowd had recoiled, but now gathered again. Voices shouted, "Where's the keeper?" "Send for a policeman!" "That brute should be arrested!"

The head-keeper was aroused by the noise. He went quickly, sensing mischief. "What's up?" he shouted, and a number of answers were volunteered. "Jinny's hurt," was the only clear one. And then a small boy said excitedly: "I seen him do it. It

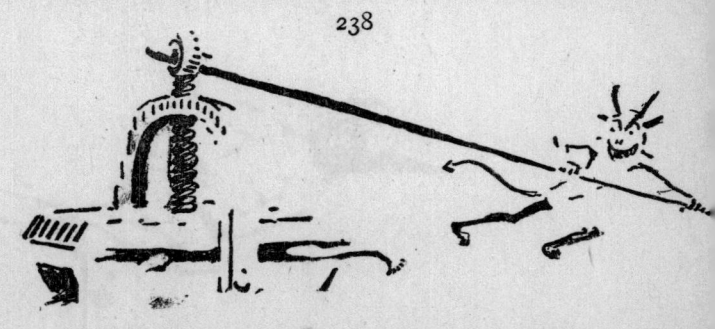

Jinny. The Taming of a Bad Monkey

was that there big feller. He stabbed her with a sword cane."

But the big fellow had disappeared. It was just as well, for the head-keeper was furious when he heard that the victim was his favorite, and if he had caught that human brute there might have been another very unpleasant scene, and equally unprofitable.

Jinny was moaning in the back of the cage. The regular keeper had tried to help, but all her old-time ferocity seemed aroused. He did not dare to come near. As Bonamy hurried to the door, the boss arrived and protested. "Now I advise you not to go in, she's dangerous. You know what her temper is." Yes, Bonamy knew better than any of them, but he entered.

There in the far corner was Jinny, holding her hand on the wounded side, moaning a little and glaring defiance at all, much as she used to do in the early days. She snorted savagely as he came near, but he stooped down and talked to her. "Now, Jinny, now, Jinny! I want to help you! Don't you know me, Jinny?"

At length he prevailed so far that she allowed him to lift her hands and examine the wound, not big but deep and painful. He washed it with antiseptic and put on a sticking plaster. She moaned while

Jinny. The Taming of a Bad Monkey

he worked, then seemed quiet. When he left she called him back in monkey fashion, a whining "errr, errr," but he was obliged to go to his office.

Next morning she was no better, and had pulled off the sticking plaster. He scolded her. "You bad Jinny," he repeated. She hid her eyes behind her arm and allowed him to put on another sticker, but she began to pull that off as soon as his back was turned, and again was scolded till she seemed ashamed, or afraid. Still it was off when next he went to the cage.

Twice a day he went to see her now, and she kept on just the same, sitting moaning in the back of the cage with her hand on the place. She always brightened up when he came in and gave that little whining "errr, errr" when he touched her. But her wound did not heal: it looked swollen, raw, and angry; and each day she was more upset when he left her. Then it got to be too much of a scene; she clung to him and kept moaning and, in monkey fashion, begging him to stay. But she would not let any one else come near, and he did not know how to fit it in with his other work. So one day he took the short cut. The boss said he was "crazy," but he did it. He took that Monkey up in his arms, and she hung around his neck like a child as he carried her to his office. She sat up in a chair

Jinny. The Taming of a Bad Monkey

and seemed quite bright, holding to the shawl he muffled around her and watching him all the time at his desk. Once in a while she would moan out that whining "errr, errr." Then he would reach out his hand and stroke her head. This pleased her, and she would give one or two little petted grunts and settle down.

But he had an unpleasant scene to face every time he had to leave the office on business. It made him feel so guilty that he transferred all the outside work he could. It was very awkward, but he could see now that Jinny wouldn't last long, and he had got so fond of her that he could not bear to cross her.

Mealtimes were making three breaks a day, which meant three upsets, so he had his food sent to him on a tray.

In a few days it was clear that Jinny was dying. She could not sit up now, her brown eyes no longer watched the clock that seemed alive, nor brightened when he spoke to her. So he swung for her a little hammock near his desk. In that she would lie and watch him with a wistful look on her face, and call him when he seemed to forget her presence. Then he would give the hammock a little swing that pleased her. He had to keep the books; she did not like to see him doing that; it prevented him looking at her. So he used to lay his left hand on

Jinny. The Taming of a Bad Monkey

her head as he worked with the right. She would hold one of her hands on her wound and tightly grasp his with the other.

One night he had given her the little soup she would take, had tucked her in her hammock as usual, and was about to leave, but she moaned and seemed to feel terribly about being left. She uttered over and over that soft, "errr, errr," so that he finally sent for some blankets and made up his mind to stay with her. But he did not have a chance to sleep. About nine o'clock she was feebly holding one of his hands in her own, and he was trying to check up some accounts with the other, when she began calling in her whining voice, but low and softly now, for she was very weak.

He spoke to her, and she had his hand, but that was not enough. She wanted something more. So he bent over her, saying, "What is it Jinny?" and stroked her gently. She took both his hands in hers, clutched them to her breast with convulsive strength, shivered all over, then lay limp and still, and he knew that Jinny was dead.

· · · · · · ·

He was a big strong man. Men called him "rough," but the tears streamed down his face as he told me the story, and added: "I buried her in

Jinny. The Taming of a Bad Monkey

the little corner lot that we keep for the real pets, on a stake at the head I nailed a smooth teak board for a memorial tablet, and on it wrote: "Jinny—the best Monkey I ever knew." As I finished writing this I found I had used a part of the cage she came in; and there on the back of it still, in large letters describing little Jinny, was the label, "*Dangerous!*"

THE END